Critical Essays on the Privatization Experience

International Papers in Political Economy Series

Series Editors: Philip Arestis and Malcolm Sawyer

This is the fourth volume of the new series of *International Papers in Political Economy (IPPE)*. The new series will consist of an annual volume with four to five papers on a single theme. The objective of the *IPPE* will continue to be the publication of papers dealing with important topics within the broad framework of Political Economy.

The original series of *International Papers in Political Economy* started in 1993 and has been published in the form of three issues a year with each issue containing a single extensive paper. Information on the old series and back copies can be obtained from Professor Malcolm Sawyer at the University of Leeds (e-mail: mcs@lubs.leeds.ac.uk)

Titles include:

Philip Arestis and Malcolm Sawyer (*editors*)
ALTERNATIVE PERSPECTIVES ON ECONOMIC POLICIES IN THE EUROPEAN UNION

FINANCIAL LIBERALIZATION
Beyond Orthodox Concerns

POLITICAL ECONOMY OF LATIN AMERICA
Recent Economic Performance

CRITICAL ESSAYS ON THE PRIVATIZATION EXPERIENCE

International Papers in Political Economy

Series Standing Order ISBN 978-1-4039-9936-8

You can receive future titles in this series as they are published by placing a standing order. Please contact your bookseller or, in case of difficulty, write to us at the address below with your name and address, the title of the series and one of the ISBNs quoted above.

Customer Services Department, Macmillan Distribution Ltd, Houndmills, Basingstoke, Hampshire RG21 6XS, England

Critical Essays on the Privatization Experience

Edited By

Philip Arestis and Malcolm Sawyer

Selection and editorial matter © Philip Arestis and Malcolm Sawyer 2009
Individual chapters © Contributors 2009

All rights reserved. No reproduction, copy or transmission of this
publication may be made without written permission.

No portion of this publication may be reproduced, copied or transmitted
save with written permission or in accordance with the provisions of the
Copyright, Designs and Patents Act 1988, or under the terms of any licence
permitting limited copying issued by the Copyright Licensing Agency,
Saffron House, 6-10 Kirby Street, London EC1N 8TS.

Any person who does any unauthorized act in relation to this publication
may be liable to criminal prosecution and civil claims for damages.

The authors have asserted their rights to be identified as the authors
of this work in accordance with the Copyright, Designs and Patents Act 1988.

First published 2009 by
PALGRAVE MACMILLAN

Palgrave Macmillan in the UK is an imprint of Macmillan Publishers Limited,
registered in England, company number 785998, of Houndmills, Basingstoke,
Hampshire RG21 6XS.

Palgrave Macmillan in the US is a division of St Martin's Press LLC,
175 Fifth Avenue, New York, NY 10010.

Palgrave Macmillan is the global academic imprint of the above companies
and has companies and representatives throughout the world.

Palgrave® and Macmillan® are registered trademarks in the United States,
the United Kingdom, Europe and other countries.

ISBN-13: 978–0–230–22252–6   hardback
ISBN-10: 0–230–22252–8   hardback

This book is printed on paper suitable for recycling and made from fully
managed and sustained forest sources. Logging, pulping and manufacturing
processes are expected to conform to the environmental regulations of the
country of origin.

A catalogue record for this book is available from the British Library.

Library of Congress Cataloging-in-Publication Data

Critical essays on the privatisation experience / [edited by] Philip Arestis
   and Malcolm Sawyer.
       p. cm. – (International papers in political economy)
   Includes index.
   ISBN 978–0–230–22252–6
   1. Privatization—Great Britain—Case studies. 2. Privatization—
   Case studies. I. Arestis, Philip, 1941– II. Sawyer, Malcolm C.
   HD4145.C75 2009
   338.941′05—dc22   2008034838

10 9 8 7 6 5 4 3 2 1
18 17 16 15 14 13 12 11 10 09

Printed and bound in Great Britain by
CPI Antony Rowe, Chippenham and Eastbourne

Contents

List of Tables vii

List of Figures ix

Preface x

List of Contributors xi

1 The Political Economy of the Private Finance Initiative 1
   Jean Shaoul

2 Private Finance Initiative and Public Private Partnerships: The Key Issues 39
   Malcolm Sawyer

3 Water Privatization 75
   David Hall and Emanuele Lobina

4 The Electricity Industry Reform Paradigm in the European Union: Testing the Impact on Consumers 121
   Carlo V. Fiorio, Massimo Florio and Raffaele Doronzo

5 Privatization and Deregulation of the European Electricity Sector 160
   Catalina Gálvez, Ana González and Roberto Velasco

6 Privatizations in Latin America 202
   Gregorio Vidal

Index 247

List of Tables

2.1	Private Finance Initiative: estimated capital spending by the private sector (signed deals)	44
2.2	Estimated payments under PFI contracts – March 2007	45
3.1	Types of water system and connections to the system in France	79
3.2	Opposition to privatization of water: some world-wide examples, 1994–2002	87
3.3	France and UK: water company ownership, December 2007	91
3.4	Renationalization and remunicipilization of water services in South America, 2007	94
3.5	Investment under-performance by Aguas Argentinas, 1993–98	96
3.6	Population connected to water and sewerage services in Buenos Aires by new extensions to the system: Projected and actual, May 1993–December 1998	97
3.7	Observed problems with BOT contracts	100
3.8	Households connected to water and sewerage by decade, Latin America	101
3.9	Investment level and growth rate before and after privatization in England and Wales	103
3.10	Selected ADB water indicators for 18 Asian cities	113
4.1	The EU-15 electricity liberalization laws	132
4.2	Timetable of the liberalization process in EU-15	133
4.3	Some regulatory indices about the electricity industry	139
4.4	Fixed-effect panel estimation	141
4.5	Estimation of a dynamic panel	142
4.6	Descriptive statistics for electricity by year and pooled sample	145
4.7	Average partial effect of consumers' satisfaction with electricity prices	148
4.8	Average partial effect of consumers' satisfaction with electricity quality	150
5.1	Production, consumption and net imports of electricity, 1990, 2004	165
5.2	Indicators relevant to market design, 2005	170

5.3	Spot traded volumes as a percentage of national electricity consumption (June 2004–May 2005)	174
5.4	Traded volumes in futures/forward contracts as a percentage of national electricity consumption (June 2004–May 2005)	174
5.5	Indicators of structural market, 2005	176
5.6	Electricity prices for households and industrial (all taxes excluded)	180
5.7	Largest European electricity companies, 2006	181
5.8	Presence of the largest companies in other Member States	182
5.9	Germany: Generation market shares (%)	188
6.1	Latin America privatization revenues, 1990–2005	211
6.2	Latin America countries: privatization revenues/GDP, 1990–2005 (%)	214
6.3	Brazil's ten largest privatizations with foreign investors, 1987–2005	216
6.4	Argentina's largest privatizations with foreign investors, 1987–2005	217
6.5	Highlights in Latin American privatizations	219
6.6	Latin America: private investment in strategic and basic public services	221
6.7	Latin America: private investment in public services	225
A6.A	Latin America largest privatizations with foreign investors, 1987–2005	240

List of Figures

3.1	Public ownership of water systems in US cities 1830–1924	78
3.2	Public and private ownership of water operators, major cities, 2006	80
3.3	Multinational water companies in 2002	83
3.4	Returns on infrastructure investment in developing countries	86
3.5	Popular support for public ownership of water industry in the UK, 2006	88
3.6	Water and sewerage connections, Latin America, 1960–2000	101
3.7	Development bank finances for public and private water projects in central and eastern Europe 1990–2002	106
3.8	Price of household water in France, 2004	108
3.9	Components of the average household bill in England and Wales 1991–2004	109
4.1	Electricity (log) price dynamics for households	138
4.2	Electricity (log) price dynamics for households	138
4.3	Trends of mean regulatory indices across EU-15	139
5.1	Electricity production as per its generation sources (%), EU-25 (1990, 2004)	163
5.2	Electricity consumption as per activity, EU-25 (1990, 2004)	164
5.3	Electricity prices for households (all taxes excluded)	179
5.4	Electricity prices for industrial users (all taxes excluded)	179
6.1	Privatizations in developing countries, 1990–99	213

Preface

This is the fourth volume of the new series of *International Papers in Political Economy* (*IPPE*). The new series will consist of an annual volume with five to six papers on a single theme. The objective of the *IPPE* will continue to be the publication of papers dealing with important topics within the broad framework of Political Economy.

The original series of *International Papers in Political Economy* started in 1993 and has been published in the form of three issues a year with each issue containing a single extensive paper. Information on the old series and back copies can be obtained from Professor Malcolm Sawyer at the University of Leeds (e-mail: mcs@lubs.leeds.ac.uk).

The theme of this fourth volume of six papers is critical appraisal of experiences of privatization and of private finance initiatives. The papers in this volume were presented at the 4th International Conference Developments in Economic Theory and Policy held at Universidad del Pais Vasco, Bilbao, Spain, 5–6 July, 2007.

List of Contributors

Raffaele Doronzo is at the Department of Economics, Business and Statistics at the University of Milan, Italy.

Carlo V. Fiorio is at the Department of Economics, Business and Statistics at the University of Milan, Italy.

Massimo Florio is at the Department of Economics, Business and Statistics at the University of Milan, Italy.

Catalina Gálvez is at the Universidad del Pais Vasco, Spain.

Ana González is at the Universidad del Pais Vasco, Spain.

David Hall is Director of the Public Services International Research Unit, Business School, University of Greenwich, UK.

Emanuele Lobina is Principal Lecturer at the Public Services International Research Unit, Business School, University of Greenwich.

Malcolm Sawyer is at the Leeds University Business School.

Jean Shaoul is Professor of Public Accountability at the University of Manchester, UK.

Roberto Velasco is at the Universidad del Pais Vasco, Spain.

Gregorio Vidal is at the Department of Economics, Metropolitan University, Iztapalapa Campus Mexico City, Mexico.

1
The Political Economy of the Private Finance Initiative

Jean Shaoul*

University of Manchester, UK

> 'PPPs are dependent on a fourth "P" politics'
>
> DLA Group (2004)

Abstract

This chapter focuses on one of the more recent forms of privatizations within the UK: the use of private finance for investment in public services via the Private Finance Initiative (PFI) and Public Private Partnerships (PPP). The turn to private finance would provide additional finance and deliver value for money through the greater efficiency of and the transfer of risk and costs to the private sector. The evidence rebuts the UK government's case for private finance. It shows that the policy, never popular, was introduced, developed and controlled by financial consultants. The policy benefits the financial elite not the broad mass of the population and, like privatization, is now being exported all over the world. While the chapter locates the policy within the development of the neo-liberal agenda, it explains its origins not just as a policy shift but as the response of the business and financial elite to fundamental changes within the workings of the economic system itself and the crisis of the profit system as a whole. The rise of the New Right was further facilitated by the failure of the old labour organizations to launch a political offensive against the neo-liberal agenda, due to its own opposition to socialism and internationalism.

Keywords: Private finance, Partnerships, PFI, PPP, Neo-liberalism, Marxism, Political economy

Journal of Economic Literature classification: L33

Glossary

DBFO	Design Build Finance and Operate
EU	European Union
FoI	Freedom of Information
NAO	National Audit Office
NATS	National Air Traffic Services
NPC	Net Present Cost
PAC	Public Accounts Committee
PFI	Private Finance Initiative
PPP	Public Private Partnership
PSC	Public Sector Comparator
PUK	Partnerships UK
SPV	Special Purpose Vehicle
TNC	Trans-national corporation
VFM	Value for Money
UK	United Kingdom

Introduction

In 1992, following the privatization of the state-owned trading enterprises in the 1980s, the UK's Conservative government introduced the Private Finance Initiative (PFI), presenting it as a way of procuring services and their underlying assets that the state could not otherwise afford. Under PFI, the private sector designs, builds and finances much needed new hospitals, schools, roads, prisons and other social and public infrastructure and provides the ancillary, but not the 'core' professional, services for 20–35 years, in return for an annual fee that covers both the capital cost of the asset and service delivery. In effect, the public authorities lease the infrastructure required for the delivery of public services from the private sector. PFI thereby provides the policy mechanism for the private sector to take over the running of public services that could not be privatized for political or financial reasons – they were not cash generative enough.

In 1997 the incoming Labour government rebranded the policy as Public Private Partnerships (PPPs) and extended it to include: a concession or franchise that charges the public directly for using the service; a joint venture between the public and private sectors; and various other hybrid forms such as sale and lease back of property, to name but a few variants. While PPPs denote joint ownership and PFI contractual arrangements, some schemes fall under neither term and the terms are used interchangeably (Treasury, 2003a). Although the finance largely comes from the private sector, the funding to pay for it may come from either the public sector or users. The Partnerships policy has, like privatization, been emulated throughout the world, where PFI may also be known as design, build, finance and operate (DBFO).

But since the government can borrow more cheaply than the private sector, it is much costlier to use the private sector as financial intermediaries. Furthermore, as well as the higher cost of private debt, there is also the cost of the profit margin of both the private partner and its extensive supply chain, and the not inconsiderable legal and financial advisers' fees to structure and negotiate the deal. In the case of the flagship London Underground PPPs, advisers' fees amounted to a staggering £500 m. Any costs incurred by private contractors on unsuccessful bids are likely to be recovered in future successful contracts, increasing the cost of subsequent PFI deals. But services that are the subject of partnership deals have never been sufficiently cash generative, if they are cash generative at all, to be run on a commercial, comprehensive and universal basis, which is why they have been provided thus far by the state. In order therefore to make such projects financially viable and attractive to the private sector the government must necessarily ensure some combination of capital grants, subsidies, implicit or explicit underwriting of the private sector's debt or the public authority's payments, bundling together of projects to increase their size relative to transaction costs, project and service downsizing, higher charges for the public authority or the users and a reduction in workers' jobs, wages and conditions. Thus, as stated in the opening quote by a legal group marketing its PFI expertise, partnerships are a political, not a market, construct that must have broader social and economic implications.

Such a policy, so fraught with definitional and financial contradictions, has necessarily proved difficult to sell to a sceptical public. Indeed, as with so many neo-liberal policies, the rationale for PFI/PPP has changed so much over time that even its proponents have described it as 'an ideological morass' (IPPR, 2001). It was originally justified as providing the capital investment that the public sector could not afford without raising taxes, breaching self- or European Union-imposed rules on fiscal deficits, the macroeconomic argument. Later, when this was exposed to be untrue, the government claimed that PFI would deliver greater value for money (VFM) over the life of the projects because the private sector is first of all more efficient and innovative than the public sector and, secondly, assumes some of the financial risks (and costs) that the public sector would otherwise carry, the microeconomic argument.

As this argument too was questioned, the government has sought to justify PFI on the basis that it delivers assets to time and budget (Treasury, 2003a).

The assumptions were very clear. First, as the private sector, incentivized by the profit motive, is more efficient and would carry many of the risks and costs previously borne by the public sector, the private provision of public services would lead to increased efficiency and wealth from which all would benefit. But the theories about the supposed superior performance of the private sector upon which such arguments are based, the property rights model associated with Alchian, Demsetz and Hayek, have been subject to little careful scrutiny as others have noted (Crain and Zardkoohi, 1978; Heald and Steel, 1986; Boardman and Vining, 1989). Most of the comparisons of public and private sector performance centred on ill-defined costs in isolation from either outputs (i.e. on economy rather than efficiency) or their social and institutional context (Spann, 1977; Pryke, 1982). Moreover, even accepting the results at face value, the forms of analysis and research methodology rarely permitted the cause of any differences to be established (Millward, 1982; Ernst, 1994). Others have based their support for such policies on public choice theory based upon the flawed behavioural assumption that the individual is simply a competitive and self-interested rational utility maximizer, so-called economic man. But again, Stretton and Orchard (1994) conclude that there is a lack of empirical support for either the theory's assumptions or predictions.

Secondly, the policy assumes that the providers of finance, the owners and financiers, are contributors to, rather than claimants on, the surplus produced by the firm. In other words, the corporation is a cooperating team of owners, financiers, managers and workers who all contribute to the creation of wealth. Such theories focus on the creation of wealth rather than its distribution. Their value is the ideological one of implying that all contribute and all should therefore get their fair share. Yet they say little about the relative contribution of each to the wealth creation process. They have still less to say about the distribution of the wealth thereby created and the competing claims on the surplus. By concentrating on the stakeholders as joint beneficiaries working in tandem, not as competing claimants on the surplus produced by the enterprise, the conflict is rendered invisible. That conflicts exist between the various stakeholders is never acknowledged, least of all the crucial one: between the workforce and the owners. This resolves the conflict in favour of the owners, since the surplus created by the workforce accrues to the owners as does any additional surplus created by the increased effort and efficiency of the workers. This is therefore a redistribution of wealth, not from the rich to the poor but from the poor to the rich.

As of July 2007 there were 590 signed PPP deals with a capital value of £53 bn (Treasury, 2007), although other Treasury sources cite larger figures. By far the largest spending department was Transport (capital value of £22 bn signed deals or 46 per cent of the total), followed by Health (£8.2 bn), Defence (£5.6 bn), the Scottish Executive and Education (each with £4.2 bn). Within transport, by the end of 2006, the capital value of signed deals for roads and bridges was about £3.2 bn. Annual payments for these projects were expected to be £6.9 bn in 2006–07, rising to £8.9 bn in 2016–17, before declining (Treasury, 2007). Total commitments for all PFI projects between 1995 and 2034, are estimated at £204 bn. However, the basis of these projections is unclear. First, the number of signed deals is lower than listed elsewhere. Secondly, it necessarily omits the new deals yet to be signed and the payments still to be negotiated for the later years of the largest scheme, the London Underground PPP. Thirdly, they are based on estimates made at financial close and are subject to upward revision in the light of inflation and contractual changes. Fourthly, the payments shown for some projects differ from the sums actually paid by the public authority. Fifthly, the Treasury elsewhere reports payments net of expected corporation tax payable (Treasury, 2003a). Thus the actual payments are likely to be very much higher and will take an ever increasing amount of the key denominator, the annually managed public expenditure that is still spent 'in house', which is itself falling due to different forms of outsourcing (Pollock *et al.*, 2001).

This chapter has two interrelated objectives. First it examines empirically the experience of one part of the Partnerships programme, the PFI or contractual model, in the UK. Using evidence derived from peer reviewed research, it seeks to synthesize the material to make an appraisal of policy in terms of the government's stated objectives. Although the evidence is largely reviewed within the British context, the analysis has international implications as PPPs are now, like privatization, being exported. Secondly, it explains the turn to a policy that displays such a disparity between the promises and the outcomes within the broader economic and political processes of the period.

The chapter starts by discussing the control and management of the PFI programme in the UK – the appraisal methodology and process – in order to understand how the assumed benefits are derived, the strength of the case for using private finance both in general and for individual projects, and the nature of the evidence required to evaluate the policy in its own terms. The next section reviews the *ex ante* evidence relating to the degree to which PFI projects are likely to deliver VFM and be affordable. The third section examines how the policy is working in terms of the stated objectives and the claimed advantages, focusing in particular on building to time and budget, robust specification, the contractual arrangements to incentivize the private sector, the financial costs in hospitals and roads, including the cost of risk transfer and affordability, risk transfer and whether it provides additional finance for investment. The fourth section shows the role of international financial consultants in promoting the policy enabling it to become so pervasive in the face of adverse evidence and public opinion. The final section locates the policy within the workings of the economic system itself and the crisis of the profit system as a whole.

Controlling the PFI programme

For a PFI project to proceed, the commissioning public authority must demonstrate that the project is likely to deliver VFM and is affordable (Treasury, 1997). This section considers each of these criteria in turn.

Value for money

VFM is dependent first upon appropriate arrangements to ensure competition for all aspects of the project, including financial advisers, to ensure that competitive pressure is exerted throughout the negotiation phase (NAO, 1997). While there was initially a high level of interest, both the size and the high cost of bidding for such contracts favoured the few large corporations. The National Audit Office (NAO, 2007) and the Public Accounts Committee (2003) have reported on the low and declining level of competition for PFI contracts. One in three PFI projects has attracted only two bidders, compared with one in six in earlier years. As projects have become larger they require and attract only a few highly experienced bidders so there is limited effective *ex ante* competition even in the best organized tendering processes (Estache and Serebrisky, 2004). Indeed, it would be highly unusual to get more than three or four bidders for large projects as industry concentration means that there are few players. For example, just six infrastructure companies won 50 per cent of the EU roads market and 16 had 90 per cent of the market (Stambrook, 2005). Concentration in the construction industry has increased in recent years following takeovers and mergers and this has led to reduced competition in PPP procurement (Stambrook, 2005). This creates increased risk for the public sector because the companies are large and powerful enough to take on the regulators in the case of conflict and force contract renegotiation on more favourable terms (Molnar, 2003). The corporations are therefore now in a position to exert the monopoly power that undermines VFM both at the point of procurement and during operations and hence control the direction of future policy in ways that privilege the few at the expense of the many.

Secondly, and this is the aspect that has attracted the most attention, the government required a highly technical appraisal methodology that emphasizes financial costs not social factors (Treasury, 1997). It specified that a modified version of the options appraisal technique should be used as the decision tool for selecting public or private finance. This entails comparing the cost of building, financing and operating the asset under private finance against the discounted whole life financial costs of a project commissioned under conventional procurement methods, known as the public sector comparator (PSC). The scheme with the lower net present cost (NPC) is assumed to offer greater VFM. The comparison also includes the costs of some of the risks associated with the construction and management of the asset and delivery of services. Since some of the risks are to be transferred to the private sector, the PSC also needs to include the cost of the risks transferred.

It is argued that the PFI option will therefore provide greater VFM than a publicly financed alternative where the public sector bears all the risks. In effect, the proponents of PFI are arguing that the difference between the cost of the public and private sector options constitutes the risk premium, which is the price the public sector is paying for risk transfer and the private sector's greater efficiency, expertise and innovation. Furthermore, such benefits outweigh the cost.

But the first point to note is that neither the appraisal methodology nor the control process is neutral. Problems of the appraisal methodology, which extend well beyond PFI, were identified many years ago in the context of investment appraisal (King, 1975). As a result, it is not widely used in the private sector because of its well-known limitations, particularly for major capital expenditure decisions because of the unreliability of financial projections that extend far into the future. It has been extensively critiqued in the context of PFI in the research literature (see for example Heald, 1997; Gaffney *et al.*, 1999a, b, c; Mayston, 1999; Pollock *et al.*, 1999; Froud, 2003; Shaoul, 2005). These authors have raised several interrelated questions: the suitability of discounted cash flow techniques in the context of public sector investment, the choice of the discount rate, the choice of the PSC, and the risk transfer that lies at the heart of the justification for PFI.

A novel feature of the methodology is the incorporation of risk transfer into the appraisal process. But the concept of risk transfer that lies at the heart of the rationale for partnerships is problematic. What is at issue is at best uncertainty, not risk, which is impossible to measure. Moreover, the risk analysis methodology is subjective and varies enormously in proportion to construction risk, the main risk, in ways that manage to just tip the balance in favour of private finance (Pollock *et al.*, 2002). Above all, it does not consider the additional costs of the risks created by PFI for the agency, other agencies and the rest of the public sector as well as the public as users (Froud, 2003).

One of the key issues was the choice of the discount rate to be used. The government's choice of a 6 per cent discount rate, higher than the pre-existing Treasury discount rate of 5 per cent that was already higher than most welfare economists believed appropriate (2–4 per cent), was 'an operational judgement reflecting, for example, concern to ensure that there is no inefficient bias against private sector supply' (Treasury, 1991, para 49). The higher discount rate favoured private finance (Gaffney *et al.*, 1999b). In other words, the choice of the discount rate was politically motivated by the desire to use private capital for public investment. Since risk was to be quantified separately and also discounted, the use of a high discount rate amounted to double counting. While the discount rate has since been reduced, the revised methodology (Treasury, 2003b) requires adjustments to the PSC that are based upon unproven assumptions and evidence that is not in the public domain.

But if the various options are so finely balanced that selection depends upon marginal differences in the discount rate or double counting via the explicit inclusion of risk, then the NPC criterion cannot reliably be used for crucial public investment decisions. It would lead to lower discounted whole life costs for the privately financed option than a publicly financed option, and hence approval to proceed with PFI projects. It is not just that such a criterion could lead to a more expensive solution that allocates resources to the private sector. Its significance lies in the fact that a greater allocation of resources to the corporate and financial sectors could in turn favour some groups at the expense of others, thereby affecting distribution and social welfare. Finally, despite the designation of the decision tool as VFM, an intuitively appealing term, what is being measured is at best the present value of the project's financial costs, its economy, not its VFM – a point the deputy controller and auditor general of the National Audit Office has also made (cited in *Financial Times* 5/6/02).

Notwithstanding these limitations, the discounted cash flow analysis with its single outcome, the net present cost, in practice provides the ultimate criterion for determining whether a project should go ahead, and social issues are largely ignored or are not operationalized. It is therefore difficult to avoid the conclusion that the government designed a system of appraisal that would provide a public justification for its policy of using private capital in public services.

Not only is the appraisal methodology skewed in favour of the private sector, so is the process and context in which the decision making occurs. Indeed, the management of the PFI procurement process is a crucial factor in ensuring that the private finance route is taken. The key government department, the Treasury, both champions and controls the PFI process. The Treasury's Projects division was initially established in 1997 with a two-year life, largely with staff on secondment from the private sector. This was later reconstituted as a Public Private Partnership, Partnerships UK (PUK), whose mission is

> to support and accelerate the delivery of infrastructure renewal, high quality public services and the efficient use of public assets through better and stronger partnerships between the public and private sectors.[1]

51 per cent of the shares are held by private sector institutions, including financial services companies that have been involved in the financing of PFI projects, and others that have PFI contracts. Furthermore, the majority of the board members come from the private sector, with the public sector represented by only two non-executive directors and the public interest represented through an Advisory Council. The structure, ownership and control of PUK are important because PUK sets the PFI agenda and reflect the conflict between policy promotion and policy control acknowledged by government (Timms, 2001).

The PFI project and the case for private finance is managed and/or vetted by the Treasury, the Departmental Private Finance Units, Partnerships UK or 4Ps (for local government projects), all of whom are staffed largely by private sector secondees from firms with a commercial interest in the policy. This means that the control process is dominated by parties that have a vested interest in the policy's expansion (Craig, 2006). Under such circumstances conflicts of interest abound.

Two of the most egregious examples of the conflict of interests, the resultant poor financial advice and the cost to the public purse, are provided by the National Air Traffic Services (NATS) and London

Underground PPPs. The NATS PPP required a government bail-out within three months of financial close in 2001. The Department of Transport had paid its advisers, one of whose tasks it was to evaluate and manage the risks to NATS' business, some £44 m. This was £17 m more than expected and at 5.5 per cent of the proceeds of the sale, among the highest of all the trade sales examined by the NAO (2002a). But despite this CSFB, the lead financial advisers, failed to evaluate the PPP correctly. It had ignored evidence and advice that did not fit with both the government's and its own desired outcome: a signed deal. CSFB told the NAO that their prime motivation was to gain valuable experience of PPPs in order to win future contracts in this new and expanding market (NAO, 2002a). The London Underground PPPs, two of whose three contracts went into administration in July 2007, were also designed by big business and consultants. Geoffrey Robinson, the millionaire businessman who had funded Gordon Brown's office while in Opposition and became Paymaster General under Brown at the Treasury, convened a group of four businessmen with experience of privatization to advise on what should be done. Their proposals were refined by the international financial consultants PWC and the law firm, Freshfields.

Finally, given the government's commitment to private finance and its refusal to make public money available for capital investment, then, as the NAO has acknowledged, these act as incentives to ensure that the case favours the private option. It is therefore almost unheard of for the PSC – developed by the public sector's private sector financial advisers – not to show that the private finance route is better VFM than a publicly financed option.

Affordability

A PFI project must also demonstrate that the annual payments are affordable. But the Treasury has not required a consistent reporting methodology that clearly describes and presents all the operating costs and enables an assessment to be made of the affordability of the scheme. Studies of PFI in hospitals have shown that affordability was indeed a problem (Gaffney et al., 1999a, b, c; Pollock et al., 1999; Froud and Shaoul, 2001). The high cost of PFI in capital terms meant that the first wave of PFI hospitals were 30 per cent smaller than the ones they replaced as Trusts adjusted their plans downwards. The affordability gap was further reduced by subsidies from the Department of Health, land sales, a shift of resources within the local healthcare economy to the PFI hospital, and 'challenging performance targets' for the Trusts' reduced workforce. Thus, PFI comes at the expense of both capacity and access to healthcare.

Several points flow from all this. First, the VFM case is necessarily based on *estimates* of future costs at the point of procurement. Second, risk transfer was conceived as the crucial element in delivering whole life economy since under PFI private sector borrowing, transactions costs and the requirements for profits necessarily generate higher costs than conventional public procurement. Thirdly, the public sector retains the ultimate responsibility for essential and often statutory services for which there is usually no alternative. That is, it retains the key political and reputational risks. This, plus government commitment to the policy, means that the private sector's revenue streams are assured, as the capital markets recognize (Standard and Poor's, 2003). The corollary is that the ability to transfer risk may in practice be very limited, and such risk that is transferred may be to the taxpayers, workforce and users rather than the private partners. Thus, the government's emphasis on VFM and risk transfer serves to disguise the high cost of PFI and downplay the importance of affordability, which in turn raises questions about VFM.

Ex ante evidence

Not only is the appraisal methodology conceptually flawed, such valuations encapsulated in VFM and set out in the projects' business cases are not generally, other than in health and education, in the public domain for reasons of 'commercial confidentiality'. This makes it difficult for the public or its representatives to make any independent assessment of a project's VFM.

While there have been numerous *a priori* reviews of PFI, there have been only a few empirical studies of the business cases that are in the public domain (see for example Gaffney and Pollock, 1999a; Gaffney *et al.*, 1999a, b; Pollock *et al.*, 2000; Pollock *et al.*, 2002). These studies show that the VFM, resting upon uncertain projections of costs far into the future, relies overwhelmingly upon estimates of the cost of 'risk transfer' to the private sector, and is at best marginal. In effect, the government created an in-built bias in favour of PFI, raising questions as to the degree to which the public agencies can and do reliably demonstrate that the higher cost of private finance is likely to deliver VFM as the National Audit Office (NAO) has acknowledged (NAO, 2000a). Such a methodology, biased in favour of the private sector option, results in a transfer of wealth from taxpayers (or users) and the workforce to the corporate and financial elite (Shaoul, 2005). However, rather than ensure a level playing field in the choice of financing options, the government's response to critical research evidence has been to dismiss the scientific evidence, discredit and intimidate critics, and ultimately exclude and ignore it (Greenaway *et al.*, 2004).

The policy's proponents studiously avoid even mentioning, let alone refuting, the evidence cited above. Indeed, while the government and the proponents of the policy claim that PFI represents VFM, they have produced little in the way of sound empirical evidence to support their case. Their evidence is largely based upon the difference between the PFI option and a hypothetical PSC, hardly an independent assessment as shown above. Apart from the London Underground PPPs (NAO, 2000a), the NAO has not carried out any assessments of projects before financial close. While the NAO has carried out numerous assessments after financial close, these were not independent in the sense that they collected new data. Nevertheless, the NAO in many cases criticized various aspects of the way the business cases were compiled and interpreted, and questioned the degree to which the projects demonstrated VFM (NAO, 1997; 1998; 1999a; 2000a).

The government for its part sought to promote its policy by commissioning several surveys of PFI, which it cites as demonstrating that PFI represents VFM. But these studies fail the first test for any evaluation: that they should be independent. They were carried out by financial consultants with a vested interest in the policy as either advisers, private sector partners in PFI deals or major subcontractors. The first report, the Andersen report, commissioned by the Treasury, is particularly important (Arthur Andersen/LSE, 2000) because it claims that PFI had 'saved' 17 per cent on the cost of conventionally procured projects. However, this is based on a sample of 29 projects (out of a possible 400 projects), whose selection is not explained. Its evidence base is the PSC used to support PFI over conventional procurement, rather than any independent analysis. But even more important, most of the savings come from just a few schemes as a result of the risk transfer to the private sector. Furthermore, about 80 per cent of these savings came from just one project, the NIRS2 project for the Benefits Agency run by Andersen's sister company, Accenture, which has become a byword for failure. In other words, the claim was largely based upon the anticipated savings of one project that were not achieved in practice. Despite this, the government has never repudiated the report, which continues to be cited uncritically.

The second report, commissioned from PWC (2001), fails to provide even the most basic information that would enable the reader to assess the methodology and the value of the findings. It is based on the perceptions of senior managers responsible for commissioning 27 PFI schemes, not users, staff or project managers. While the report does not explain the sample choice or even provide any evidence about the nature or sector of the schemes, its author explained to this writer that PWC largely selected projects with which PWC had been involved as adviser to either the public or private sector, excluded IT projects and included the first eight DBFO road schemes.[2] The report does not contain any supporting financial or other empirical data on service or volume levels.

A third widely cited report, authored by the Institute of Public Policy Research (IPPR, 2001), the think tank with close relations with the Labour government, was sponsored by KPMG and other private sector companies with a vested interest in the use of private finance. It too used secondary, *ex ante* evidence. While the report had reservations about the use of PFI in health and education, it did endorse the turn to private finance via partnerships.

This review shows first that the government provided no empirical evidence to support its policy choice. Secondly, its choice of private finance for specific projects was based upon a flawed methodology and a process biased in favour of private finance, and rested upon the risk to be transferred to the private sector. Thirdly, empirical studies by academic researchers show that VFM has not been demonstrated. Indeed, under conditions where partnerships are the only means available to the public sector for procuring goods and services, then the VFM case is little more than a rationalization for a decision already taken elsewhere. Fourthly, since the surveys commissioned by the government are neither independent nor methodologically rigorous, little reliance can be placed upon them. Finally, any evaluation needs to consider the outcomes in terms of the stated objectives of the policy, with a particular emphasis on the actual cost of private finance, the risk premium and the benefits attributable to the cost of the risk premium.

Ex post evidence

There has as yet been little in the way of financial evidence as to how the turn to private finance is working out in practice. Indeed, Hodge's review of Australia's experience (2005) notes that there has been no comprehensive evaluation of PPPs; parliamentary enquiries have revealed 'a paucity of quantitative information relating to risk experience and weak financial evaluations' of the comparative performance of PPP and traditional mechanisms; and therefore that 'much of the political promise has not yet been delivered'.

In the absence of either a comprehensive evaluation of such claims or systematic evidence in the public domain that would enable such claims to be evaluated, the evidence presented here about how PFI is working in practice in relation to the claims used to justify private finance is drawn from a wide variety of both primary and secondary sources. These include NAO reports and academic, corporate and other commentaries.

Building to time and budget

The government claims that in contrast to conventional public procurement, PFI projects have been built to budget and on time. But first of all this assumes that public procurement has been consistently late and over budget, and that this is greater than in the private sector. Good evidence on this is lacking, in part at least because so little was commissioned by the public sector after 1976. In the case of the NHS, cost overruns on the price agreed at financial close on conventional procurement in the early 1990s were of the order of 8 per cent. Secondly, there are indeed well publicized examples of huge cost and/or time overruns on major projects, including the British Library, the Jubilee Line, and the Scottish Executive building, to name but a few. But similar examples can be given of such cost and time overruns in the private sector, such as the new Wembley Stadium. The most egregious example is the delay and escalation in cost of the upgrade of the West Coast Main Line which rose from an estimate of £2.5 bn to £13 bn under the privatized Railtrack, before being reined back to about £7.5 bn by Railtrack's all but renationalized successor, Network Rail (NAO, 2006a). In the context of PPPs, the now failed Metronet contracts for the London Underground PPPs, the largest PPP deals ever signed, are both late and over budget by nearly £1 bn just four years into operation. Thirdly, as Flyvbjerg *et al.* (2003) have pointed out, cost overruns are a common phenomenon in high profile or megaprojects where political reputations and legacies are involved, whether publicly or privately financed. This is because everyone involved has an incentive to ensure that costs are underestimated and revenues inflated to ensure that the project gets the go ahead to proceed.

The government's case for building to time and budget under PFI rests upon two reports by the NAO (2001; 2003a), which were surveys and consultations with project managers, and were not backed up with any data on cost and time overruns or comparisons with conventional procurement, another study cited by the NAO (Agile, 1999) and a Treasury report (2003a), both of which contained neither data nor methodology. As Pollock *et al.* (2007) have shown, a fifth report (Mott Macdonald, 2002) contained so many flaws in the study design and methodology that the results are uninterpretable.

While the NAO reported that the aims of PFI had generally been met in the construction and design of the 11 hospitals built to date, this must be qualified by the widespread criticism of at least one hospital (it has corridors too narrow to permit more than one trolley) and problems in other hospitals. Other more strategic criticisms have been made of their design (Appleby and Coote, 2002; Worthington, 2002). In the context of schools, the Audit Commission's review of PFI schools (2003) found that PFI did not guarantee better buildings despite their higher cost. All this ignores the extent to which costs escalate during procurement, as others have shown in the context of new PFI hospital builds (Pollock et al., 2007). In the case of criminal justice contracts, court service projects have escalated in price, refuting the claim that PFI contracts deliver fixed prices (Centre for Public Services, 2002).

Furthermore, it should also be noted that over the full planning period of a project the time taken for selection, bidding and contract negotiation processes under PFI may be months, or even years, longer than for Exchequer financed schemes, introducing delays and costs to the procurement process (NAO, 2007). In other words, understanding the reality that underpins the rhetoric of 'on time and to budget' is not straightforward. It needs to be understood in the context of the costs of this achievement over the full planning period and not just the time period between financial close and project construction. The (high) costs associated with bidding have already resulted in fewer competing bids, and recouped or reimbursed costs for failed bids provide no VFM. In essence, it is difficult to quantify the benefit of finishing on time and to assess this against the increase in price that the contractor demands to carry the risk of timely completion, a cost that is shown below to be a high one. However, if this balance is a positive one, then such benefits are not exclusive to PFI, but could also be achieved with similar contractual arrangements for conventionally financed projects. Furthermore, these issues need to be considered in a holistic evaluation of PFI rather than in the context of individual projects.

Robust specification

While the Treasury (2003a) argues that there will be greater discipline at decision making about what the public sector is procuring and that the independent due diligence carried out by the financiers of the project will ensure a robust project specification, this has not always turned out to be the case. The Channel Tunnel Rail Link had to be renegotiated within months of signing. The National Air Traffic Services PPP collapsed within three months of financial close for reasons that were entirely foreseeable despite the official line that it was due to the collapse in transatlantic flights after the terrorist bombing of the World Trade Centre in 2001 (Shaoul, 2003). The Royal Armouries Museum deal had also to be bailed out, and the QEII Greenwich hospital trust is technically insolvent (PWC, 2005), which is in part at least due to the £9 m extra costs resulting from PFI. Most recently Metronet, which was responsible for two of the London Underground PPPs, has collapsed with £2 bn of debts guaranteed by Transport *for* London.

This is not just a British phenomenon. Estache and Serebrisky (2004), in their overview of transport PPPs, note that such projects have not been uniformly successful. With a high cost of capital and lower than expected demand, 55 per cent of all transport concessions implemented between 1985 and 2000 in Latin America and the Caribbean had to be renegotiated, and that such renegotiations took place within about three years. While governments gained in the short term from any proceeds and the low level of public investment, the renegotiations led to higher expenditure via up front capital grants, subsidies and explicit debt guarantees to the private sector to make the schemes viable. In Mexico, new toll roads were unsuccessful and had to be taken back into public ownership.

Boardman *et al.* (2005), in their review of private toll road cases in North America, report that even after refinancing and gaining tax exempt status and extra ridership, the Dulles Greenway project was still making heavy losses. In the case of the Highway 407 Expressway, the Ontario provincial government had to assume the financing of a cost it had sought to transfer to the private sector in order to make the road affordable to users. In the context of Spain, which has by far the longest experience of private finance in roads, three schemes had to be taken into public ownership in 1984, a large number of the foreign loans had to be renegotiated, state loans were made available, the remaining contracts had to be renegotiated and in some cases, public subsidies were given (Farrell, 1997). Hungary's M5 project had to be restructured within months of signing. In the case of the M6 toll road in Britain, where traffic flows are much lower than forecast and the concessionaire is unable to break even, this has led to the concessionaire lobbying for development in the region to promote traffic growth and offering to pay for a new link road that will bring traffic to its toll road.

In short, the claims for robust project specification have not always been realized. At the very least, the robustness has served the banks, not the public sector, which to date have not lost out when projects have failed.

Penalties to incentivize operational performance

For several reasons it is difficult to know the degree to which the penalty and incentive system operates to ensure satisfactory delivery of contracted services. First, the size of the penalties relative to the baseline payment below which the total payment cannot fall is not generally disclosed. One hospital for example reported that maximum deduction for poor service delivery was £100 000 on expected annual payments of £15 m (Edwards et al., 2004), which provides little effective sanction. Anecdotal evidence confirms that elsewhere the maximum penalties, while larger, are indeed small relative to the annual payments. Secondly, the public agencies neither report the standards of performance nor the amount deducted for poor performance

Thirdly, there have been numerous adverse press reports in the UK of poor service delivery in hospitals under the contract, some of which are documented in evidence to the Health Select Committee (2002) and similar concerns about poor performance in schools projects. Metronet, the private sector partner for two of the three London Underground PPPs, was heavily criticized by London Transport, the Office of Rail Regulation and the Arbiter for failing to meet the targets set for investment and maintenance and overspending by nearly £1 bn in its first 7.5-year contract, owing to 'not working economically, efficiently or in line with industry best practice'. Nevertheless, according to the credit ratings agency Standard and Poor's (2003), there have been few deductions on PFI/PPP contracts and these have been small, in part at least because of the complexity of the contracts that have proved difficult in practice to enforce. Recourse to legal action is beyond the financial and human resources of most public agencies. Since in many cases the original contract negotiation team has moved on, it is difficult to know the assumptions and intentions underlying the contract, or even whether such contractual (c)omissions were by default or design.

Furthermore, not only is performance not always satisfactory and penalties small, the monitoring of contract performance has turned out to be much more costly than anticipated and performance indicators have been difficult to operationalize, because of the subjective nature of the outcomes. In many cases, the performance data is produced and owned by the contractors, meaning that the public agency has had to validate the data and/or collect their own in order to verify payments against performance (Edwards et al., 2004). Contract changes have been time consuming and complex.

The NAO, in its investigation into PFI prison performance, reported that operational performance against contract had been mixed (NAO,

2003b). HMP Altcourse at Fazakerley, the first PFI prison, was controversial from the start because of its poor planning, lack of scrutiny of costs, a flawed savings assessment, operational performance failures and lastly the refinancing scandal that saw the private sector refinance the deal in a way that generated extra £11 m for itself while at the same time increasing the risk to the public sector (NAO, 2000b). A report on prison performance noted that prisoners were confined to their rooms for longer periods and that their cells contained 'substantial ligature points' that 'rendered the cells unfit for use at all' (Chief Inspector of Prisons, 2000). But PFI contracts, even when 'successful', have hidden costs to the rest of the public sector. The Centre for Public Services (2002) found that the private sector paid lower wages to its prison staff than did the public sector and some of its workforce were paid such low wages that they qualified for working family tax credits, in effect a low wage subvention by the state to the private sector.

As is almost universally accepted, operational performance has been conspicuously poor in IT projects, where the payment mechanisms clearly failed to incentivize the contractor. Even where penalties could have been invoked, these were waived in the interest of good partnership working and/or not jeopardizing the policy, as in the case of the Passport Agency (NAO, 1999b) and NIRS2 projects (Edwards and Shaoul, 2003). Indeed, the outcomes of IT projects in the Benefits recording and payments systems, the criminal justice system and other administrative services have been so poor that even the government has had to admit that PFI may not be the best means of procuring IT services (Treasury, 2003a) and to abandon PFI for IT projects.

Thus, once again, understanding the reality that underpins the rhetoric of 'incentivizing the private sector' is not straightforward. Such evidence as exists suggests the scale of the penalties, the complexity of the contracts and the relative power of the partners do not provide the incentives that PFI's proponents claim, while simultaneously imposing additional costs on the public sector to ensure contractual compliance.

The financial cost of PFI

There have been few studies that produce systematic financial evidence about the cost of PFI projects once they are operational. This section cites two, one in hospitals and the other in roads.

(i) Hospitals

The first twelve operational PFI hospitals in England as of 2001 had capital costs of about £1.2 bn, combined annual PFI payments of about

£260 m in 2005, and total expected payments of about £6 bn over the 30-year life of the projects (Health Select Committee, 2000). But the actual payments to the private sector turned out to be 20 per cent higher on average than originally estimated: payments were as much as much as 71 per cent higher for North Durham, 60 per cent for South Manchester and 53 per cent for Bromley (Shaoul et al., 2008). While this may be due to volume increases, inflation, contract changes and failure to identify and/or specify the requirements in sufficient detail, e.g., the failure to specify marmalade for patients' breakfast led to an increased charge, such contract drift suggests, at the very least, that there will be further increases. Extrapolating from the 2005 costs, which are likely to increase further, the total cost of PFI for these Trusts is therefore likely to be £7.2 bn, very much more than the £6 bn predicted at financial close, and reflected in the Treasury's estimates of future commitments.

The hospital Trusts' PFI charges, including both the availability and service elements, took on average 12 per cent of income in 2005. The case of Dartford is particularly interesting because even after a refinancing deal that led to a reduction in their charges, PFI charges still took 17 per cent of income. While the Trusts received a 56 per cent increase in funding (adjusted for any mergers) as part of a wider policy of more funding for the NHS to cover additional services, investment, salary increases and, in some cases, a specific increase to cover some of the extra costs of PFI, PFI charges were still taking the same proportion of income, raising questions about the affordability of PFI. It is therefore difficult to avoid the conclusion that without the increase in funding, PFI was unaffordable.

Despite the increase in funding, the Trusts' financial situation, like many non-PFI Trusts, was neither stable nor robust. Without a detailed study of each Trust's caseload, it is difficult to determine the role of PFI as other factors have intervened. But two examples illustrate some of the problems. In the case of South Manchester, which had suffered a £7 m deficit in 2003, this was because it was unable to shift a £20 m caseload to other hospitals that had been part of a wider reconfiguration underpinning the original business case, thereby triggering volume increases in its PFI charges. The QEII Greenwich Trust, with one of the largest deficits – £9.2 m in 2005 – declared that it was technically insolvent and was locked into a PFI deal that added £9 m to its annual costs over and above that built under conventional public procurement (PWC, 2005). Without government support, its long term financial situation was insoluble.

Irrespective of any causal role in the Trusts' financial problems, PFI charges constitute a 'fixed cost' that cannot be reduced and are significant when margins are low because of other rising costs. This serves to reduce their flexibility in managing their budgets which must create affordability problems when the Trusts have always struggled to break even.

The private sector companies, special purpose vehicles (SPV) or consortia organized as brass plate companies, operate in a complex and opaque web of subcontracting to their sister companies that increases the costs and complexity of monitoring and enforcing the contract, and makes it impossible to assess the parent companies' costs and total returns. They spent £140 m in 2005 on subcontracting, presumably for operating and maintaining the hospitals, the service element of the unitary charge, an amount that is likely to rise as the hospitals age. After paying interest on their debt, which was higher than the total construction cost and rising, of about 7–8 per cent, the SPVs reported a post tax return on shareholders' funds in excess of 58 per cent in 2005, after negative returns in the early years. The SPVs' high effective cost of capital (£123 m in 2005) means that the annual risk premium, the difference between public and private sector interest as defined by the NAO (1998), was £55 m, equivalent to 20–25 per cent of income received from the Trusts. In other words, PFI is costing an extra £55 m a year over and above public debt, assumed here to be 4.5 per cent, the prevailing rate on Treasury gilts. All this raises questions about the affordability of PFI in practice and future service provision, an issue which the emphasis on VFM downplays.

But this analysis underestimates the total cost of private finance, since the SPVs' parent companies have additional, undisclosed sources of profit that constitute leakages from the public purse. These include the profits from subcontracting the construction, operation, maintenance and financing of the projects to related companies, direct user and staff charges for parking, canteens and telephone/television which also represent lost income to the Trusts, the proceeds of land sales and any refinancing of the SPVs' loans. None of these are possible to quantify, if indeed they can be identified, in a systematic way, making it impossible to establish the total cost of using private finance.

(ii) Roads

While PFI in hospitals has been hotly debated owing to the availability of financial evidence, PFI in roads, where it is known as DBFO, has largely been invisible as the business cases used to support the use of private finance have not been in the public domain. Private finance in roads has been deemed a 'success' by the government and the Highways

Agency alike. However, this was and is a consequence of very high payments to the private sector. Shaoul et al. (2006) examined the first eight DBFO contracts paid for on the basis of shadow tolls and signed by the Highways Agency. They found that the projects are costing about £220 m a year or £6 bn over 30 years. Payments in just three years for which information is publicly available were £618 m, more than the £590 m cost of construction, refuting the claim that the government could not afford the capital cost.

After paying interest on their debt, which was higher than the total construction cost, of about 9 per cent, the SPVs reported a post tax return on shareholders' funds of 29 per cent in 2002. The annual risk premium was about £56 m or about half the cost of capital (£103 m) and one-third of the income received from the Agency in 2002. With annual operation and maintenance costs of about £50–60 m a year, or £1.8 bn in total, this means that after paying interest on debt (about £1.8 bn), itself more expensive than public debt, the Agency is paying nearly £1.8 bn (out of a total of £6 bn) for the major maintenance and private sector profits, a high price for risk transfer. Thus 'success' comes at the expense of affordability and must entail service cuts elsewhere. Indeed, a Highways Agency official has admitted that annual payments for all its contracts are £300 m a year, or 20 per cent of its budget for 8 per cent of its roads. The contract for the M25 will add a further £300 m a year, meaning that 40 per cent of the budget will be committed for a very small proportion of the network (Taylor, 2005). As in the case of hospitals, this analysis underestimates the total cost of private finance, since the parent companies have additional, undisclosed sources of profit through subcontracting the construction, operation, maintenance and financing of the projects to related companies, as well as refinancing gains, which make it impossible to establish the total cost of using private finance. But at the very least, these findings rebut the arguments what the private sector would find the finance that the public sector could not (the macroeconomic or additionality argument).

Risk transfer

Most of the additional cost of private over public finance is justified in terms of risk transfer, largely construction not operational risk, and is known as the risk premium. There is, however, no yardstick by which to measure *ex post facto* whether the annual risk premium identified at £55 m and £60 m a year for hospitals and roads respectively is a reasonable reward for the risks transferred to the private sector consortia.

But the following points are crucial to understanding risk from the private sector's perspective.

First, the SPV's parent companies invest very little of their own money in the company, typically 3–5 per cent of total capital employed by the SPV, which has no recourse to its parent companies should costs rise more than anticipated in ways that cannot be recouped from the public sector. The contract is implemented via subcontracting to subsidiaries of the parent companies, which could indeed lose money or even go out of business, as has in fact happened if they submitted over-optimistic bids. The main cost, as shown earlier, is the cost of finance. Should the SPV be put into administration or terminate its contract with the commissioning agency, then the public sector must continue to honour the financial obligations to the banks. There is therefore little direct risk to the parent companies and almost none to the banks, since the state *de facto* guarantees the debt.

Secondly, in the case of roads, it is difficult to see, given that the contracts involved roads that had already been designed and gone through all the planning stages, thereby reducing some of the main risks, how such a high 'risk premium' could be justified (Shaoul *et al.*, 2007a).

Thirdly, it is unclear why the cost of risk transfer is so high given that after completion of the construction phase, about 40 of the SPVs have been able to refinance their deals. Furthermore, these refinancing deals serve to create additional risk for the public sector (NAO, 2002b; 2005; 2006b). In the case of the Norfolk and Norwich hospital, the consortium was able to take out a larger loan over a longer period (its contract with the Trust was duly extended to accommodate it) and repay other loans, leaving a surplus for its parent companies to invest elsewhere. The Trust could therefore find itself exposed to additional termination liabilities should the contract be terminated for any reason. This increased exposure would occur when the private sector had received most of the benefits and be facing additional costs associated with long-term maintenance, thereby tempting the private sector in adverse circumstances to cut and run, as indeed has been the case with unprofitable rail franchises.

Fourthly, as well as the refinancing gains, many more companies have been able to sell their shares in PFI consortia at a considerable profit on their original investment and at several times the level of annual post tax profits. This has not only brought additional profits to the original owners in the early stage of the projects when profits were not expected to be high but also implies that their new owners envisage continuing and more attractive rates of return on their investment.

Taken together, this means that if the project is successful, then the public agency may pay more than under conventional procurement: if it is unsuccessful then the risks and costs are dispersed in unexpected ways as a study of failed IT projects has shown (Edwards and Shaoul, 2003). Although a project may fail to transfer risk and deliver VFM in the way that the public agency anticipated, the possibility of enforcing the arrangements and/or dissolving the partnership is in practice severely circumscribed for both legal and operational reasons, with the result that a public agency may be locked into a partnership for better for worse. This in turn undermines the power of the purchasing authority to incentivize its partner while strengthening the contractor's already powerful financial and monopolistic position, under circumstances where it is beyond the reach of public accountability and scrutiny.

In short, the experience shows that far from being a neutral policy-making decision tool, 'risk transfer' disguises the political and social consequences. In effect, the risk transfer is not from the state to the private sector, but from the consortia to its subcontractors and their workforce and to the public as tax payers and users, a travesty of risk transfer. The beneficiaries are the banks and to a lesser extent the consortia and parent companies.

Additionality

Since the public sector repays the full cost of private finance via annual payments spread over 30 years, it does not access new forms or higher levels of funding than would otherwise be the case with public funding. It simply spreads the cost over a longer period and ultimately costs at least three times the original outlay.

In the context of hospitals, it should be noted that, first, while the government claims that PFI has led to the largest building programme in the history of the NHS, the first wave of PFI hospitals were so costly that they created an affordability gap, leading to asset sales, extra subsidies, charity appeals and cuts of up to 30 per cent in bed provision, that is, hospital contraction (Gaffney *et al.*, 1999a). Secondly, the annual observable extra costs of private finance in hospitals, £55 m, extrapolated across the whole PFI hospital programme, shows that the £9 bn programme is costing at least £430 m extra every year, equal to at least two new hospitals every year or 60 over the life-time of the programme. The additional cost of private finance in turn creates affordability pressures for the Trusts, which have been cushioned to some extent by increased funding. Under the new funding regime where money follows patients on the basis of average costs, there will be further cost pressures for Trusts locked into PFI contracts since they have higher fixed costs than non-PFI Trusts, as the QEII Trust noted (PWC, 2005). Thus, PFI creates budget inflexibilities that increase the pressure on the NHS to cut its largest cost: staff and thus access to quality health care. In short, far from financing an expansion of the NHS, the policy has already led and will lead to a further contraction.

In the context of DBFO in UK roads, as the evidence above has shown, the £590 m construction costs were paid for in three years. This means that far from providing additionality, the new construction (and maintenance) comes at the expense of other Highways Agency projects, as the uneven ration of expenditure on PFI to the overall road network reveals (20 per cent of the budget on 8 per cent of the network).

In short, this analysis has demonstrated that the outcomes do not match the claims. This is because the government's claims ignored the competing demands of the numerous stakeholders and the particular characteristics of public services: cash strapped with no excess capacity to enable 'surplus fat' to be trimmed without affecting service delivery. In these circumstances it was and is impossible to reconcile all the conflicting claims on the funds and protect both the taxpayers and users. PFI ensures a resolution of the distributional conflict in favour of the corporations and more particularly the financial sector, who are its chief promoters, under the guise of additionality, risk transfer, efficiency, incentives, and so on. Thus while the government's case rests upon risk transfer, additional investment and private sector efficiency, and therefore benefits for all, the real effect is the redistribution of wealth to the financial and corporate sectors. The government, by focusing on a concept as ambiguous as VFM under conditions where no public finance would be made available, itself a policy choice, made the distribution issue invisible in order to justify a deeply unpopular policy.

The policy promoters

Privatization in its various forms was neither the result of a widespread movement among the public at large, nor was it popular. Governments of all political shades introduced it at the behest of big business, not the public. The creeping privatization of public services via PFI/PPP is if anything even less popular and often justified by reluctant public authorities charged with its implementation in terms of 'there is no alternative'. But as the evidence shows (Shaoul et al., 2007b), the policy formulation process is driven by private sector staff on secondment or loan to the Treasury. The procurement process is lengthy and opaque, owing to 'commercial confidentiality', and brokered and administered not by the civil service or local government officers, but by 'financial advisers', particularly the Big Four accountancy firms, both at Treasury/Department and public agency levels.

The accountancy industry, whose growth is bound up with the state-backed monopoly of the annual external audit, has become increasingly concentrated and is itself now part of the wider international business services industry. This has given the Big Four ever greater financial clout and influence over both the formulation and implementation of the policy. There were recurring dangers of conflicts of interests when they advised and later staffed government departments, in some cases free of charge. While the policy of bringing in 'advisers' began under the Thatcher government in the 1980s, this increased markedly under New Labour, particularly as it aggressively promoted PFI/PPP and changed the relationship between the higher echelons of the public sector and business. There is now a constant merry-go-round. Senior civil servants leave to take up lucrative posts in the private sector where they can put their knowledge of and contacts in government to good use. The Big Four accountants enter the civil service in general and the Treasury in particular. While they may be more altruistic than their peers in that they accept pay cuts (Gosling, 2004), they believe that private sector financial management is 'better' and has much to teach the public sector. The significant numbers entering the Treasury is important since it has become by far the most powerful government department under New Labour.

The Big Four have been able to consolidate their grip on policy, which they have a commercial interest in expanding, via their presence and positions of influence in Whitehall and PUK, and their social networks. One of the key ways they have been able to formulate and promote policy is by their reports that frame the methodologies to be used for appraising projects (ensuring that PFI is the preferred option and is suitably profitable to the financial institutions at least) and evaluate the outcomes. They have been able to ensure its implementation by their role as financial advisers to public agencies considering procurement via PFI.

This has several interrelated implications. First, active championship of PFI/PPP by the Treasury and government departments raises issues about conflicts of interest if the same public bodies both promote and control these projects, particularly where there is a lack of public finance for conventional procurement. Second, this conflict is further exacerbated when such bodies themselves reflect or are owned or controlled by vested interests, thereby jeopardizing any possibility of regulating and scrutinizing it in the broader interest of the citizenry.

Thirdly, the Partnerships policy itself blurs the boundaries between the public and private sector at a number of levels, making responsibility more and more confused and difficult to pin down. At the policy level the Treasury and civil service have become ever more closely intertwined with business. Hampered by *Freedom of Information* rules that enable government to avoid disclosure where it relates to policy formation, the general public and its representatives are unable to scrutinize decision making effectively. At the project procurement level, 'commercial confidentiality' means that the business cases used to justify private finance are rarely released, except in the case of hospitals, even after financial close. At the delivery level, it becomes ever more difficult to track and control public expenditure, which is increasingly channelled through the private sector, as others have shown (Edwards *et al.*, 2004). All this renders the traditional mechanisms of accountability for public expenditure and control of the executive obsolete.

Fourthly, it was not just union leaders such as Unison's general secretary who termed the relationship between the government and its financial advisers 'a web of deceit bordering on corruption'. Sampson (2005) found that senior accountants 'were also shocked that the government could allow such an obvious conflict of interest'. He believed that the Big Four were becoming a serious threat to democracy as their networks penetrated Whitehall. This is particularly important when the government excludes oppositional voices. While large corporations have always been able to exert power and influence, the last 10 years have seen a huge intensification of this process in which PFI/PPP has played a major role.

Fifthly, the increasing privatization of policy formation by those with very different interests to those of the public at large has in turn reinforced their political and financial position, not just at the national but the international level. The opening up of public services to the private sector is part of an ongoing process whereby the social and public services pass into the private sector through buy-outs, subcontracting, and sale and lease-back operations such as the PFI. Increasingly such services, like all the former nationalized industries, are then integrated into the wider international economy as they are taken over by the transnational corporations (TNCs).

Private finance and neo-liberalism

How is all this to be explained? If the stated reasons for PFI do not match the results it is because they are part of a very different agenda: the opening up of public and social services for private profit. Such services, previously outside the realm of private exploitation, have been transformed into commodities to be produced for profit, not use, with public service employees the source of that profit. This marks a very definite transformation of social relations in a number of important respects. First, the relations of production in the public sector are being realigned so that they match those of the private sector. Secondly, services, funded by the public through taxation, are organized by the state to serve the financial interests of private corporations not the public. Thirdly, the public is being reconstituted as the customer for the goods and services so produced. Fourthly, these changes are part of an ongoing process whereby the social and public services pass into the private sector and are then taken over by the TNCs. In other words, the social welfare functions of the nation state are being integrated into the world economy, but for the benefit of capital, not labour or the public at large. The broader significance of these neo-liberal policies is that they provide the ideology and mechanisms to accomplish an international market for health, education, transport, infrastructure, and so on.

While there have been many reviews of neo-liberal policies, there have been few attempts to explain their source and development within the workings of the economic system itself. The turn to private finance is widely viewed as the outcome of the 'free market' ideology rather than as an expression of the crisis of the profit system as a whole.

The ideological sea change, known as the New Right Agenda, that took place in the mid-1970s was and is presented as simply a policy shift that occurred during the 1970s (and could therefore by implication be reversed by another policy change). But, in fact, it reflected the response of business leaders and their governments to the objective changes that had taken place in the world economy. The downturn in the rate of return on capital employed in the 1960s and 1970s was the driving force for several interrelated processes: the globalization of production in order to reduce costs, and the development and application of new technologies of production: computers and telecommunications. Changes in technology enabled ever fewer productive units to supply a world market. Together these processes have been responsible for a transformation of the structure of the capitalist economy. The resulting global mobility of capital spelt the end of the programme of Keynesian national regulation that formed the basis of the postwar welfare state and the state-owned enterprises and services.

At the same time, the technological innovations in the production process based on the computer chip enormously intensified the crisis of the profit system. The mission of a capitalist enterprise is to make not simply an absolute level of profit but a level of profit proportional not to sales but the amount of capital employed. This is typically 10–15 per cent, and must always be higher than the prevailing interest rate, since this represents the basic return available to the providers of finance. As the amount of capital employed in an enterprise increases, so must the profit.

The cash surplus or surplus value – the basis of profit – represents in the final analysis the surplus labour extracted from the workforce. But the essence of new technology and cost cutting is the replacement of value-creating labour by capital equipment in the production process. Consequently, rather than alleviating the tendency of the rate of profit to fall, it has worked to exacerbate it, as Armstrong *et al.*'s analysis shows (1984). While this was and is largely invisible in the public debates, it was this that lay at the heart of the policy shift and the New Right Agenda. It is this falling rate of profit relative to the amount of capital employed (even though the absolute amount of profit may be rising in particular industries or companies) that lies behind the successive waves of mergers and cross-border mergers in the 1980s and 1990s: corporations sought to cut costs and sell off surplus assets, thereby reducing the amount of capital employed.

Under conditions where the overall mass of surplus value was expanding, as in the first 25 years after the Second World War, capital was able to tolerate the welfare state and even welcome the nationalization of basic industries. Such policies provided a means of containing and regulating the class struggle since the government as owner was able to pay higher wages and improve working conditions as in the mines and railways in Britain. The welfare state meant that employers did not have to make extensive provision for health insurance and retirement for its workforce as did employers in the US during the period of the long boom. Nationalization shifted the cost of investment in capital-intensive industries on to the taxpayers while enabling their former owners to reinvest the proceeds from compensation in more profitable ventures. At the same time the nationalized industries and services could be run in ways that constituted a subsidy to industry. Indeed, the nationalizations in the 1940s were justified with claims of the increased efficiency that would flow from the restructuring and increased investment that only government could provide (Millward, 1999).

But under conditions where the tendency is for the mass of available surplus value to decline, deductions in the form of corporate taxation to finance social welfare became increasingly intolerable. One of the responses of the ruling elites everywhere was to attempt to claw back a portion of the surplus value previously appropriated by the state in the form of social welfare provision to the working class. Furthermore, the 40 per cent or so of GDP that could not be cut for political reasons and did not directly yield a profit had to be opened up via privatization, PFI, 'partnerships', outsourcing, and all the rest, to private profit. The past 30 years have been characterized by an on-going assault on the social position of the working class as capital seeks to overcome the pressures on the rate of profit. But all these measures have failed to establish a new equilibrium based on the expansion of the mass of surplus value. Hence the continuous rounds of privatizations and cost cutting. No sooner is one round of cuts completed than another begins.

While there has been a tendency for the mass of surplus value to decline, there has been a huge expansion in the amount of finance capital available seeking new and profitable outlets for investment with an acceptable rate of return. This glut of global finance capital has sought ever more sources of profit and fewer restrictions imposed upon the TNCs' activities by national governments as they undertake the production of and investment in goods and services on an international scale. The TNCs seek to extend and deepen their global reach at the expense of the universal right to water, sanitation, transport, energy, health care and education and basic democratic rights. But the creeping privatization espoused by all governments, the international financial institutions, the European Union, and the TNCs, also expresses more fundamental processes – the inherent drive of the productive forces to break free of the constraints of the outmoded nation state system.

This analysis has attributed the neo-liberal policies that are so detrimental to both social equality and public services to the economic system itself, not simply to an ideological sea change that occurred in isolation from the changes in the structure of the international economy. It therefore raises important questions about the character of any oppositional movement against such control by the giant corporations. While there has been widespread opposition to the policies of the international financial institutions of global capitalism, such as the G8 group of nations and the World Trade Organization, the opposition has largely taken the form of opposition to globalization as opposed to global capitalism. Furthermore, it was unclear on what basis they were opposing global capital or globalization, whether they agreed with each other and, crucially, to whom or to which social layers they were speaking: international institutions, governments, trade and labour leaders, or the broad mass of the world's population.

In other words, there has been a basic confusion that has identified 'globalization' with 'global capitalism'. While globalization, the increasingly global character of the production and exchange of goods and services, is in itself a progressive development, the destructive consequences flow not from globalization as such but from the subordination of all economic life to the anarchic pursuit of profit, based upon national forms of political organization. Thus the real question is not how to return to some mythical golden age of national economic life, but who will control the global economy and in whose interests will it be run.

It is impossible to succeed against the restructuring of public services in the era of neo-liberalism, isolated from any wider social and political struggle in the working class. An examination of the past century shows that whatever gains were made were by-products of major political and social struggles of the working class, struggles that were led by socialists against the existing opportunist leaderships. As Rosa Luxembourg pointed out 100 years ago,

> Work for reform does not contain its own force, independent from revolution. During every historic period, work for reforms is carried on only in the direction given to it by the impetus of the last revolution, and continues as long as the impulsion of the last revolution continues to make itself felt. (Luxemburg, 1988, p. 49)

Just consider the origins of the nationalized industries. Nationalization was the product of a big movement in the working class. The British ruling class, acutely conscious of the Russian Revolution in 1917 and the revolutionary upheavals in Europe that followed the Russian Revolution and the First World War, grudgingly allowed a reformist Labour government to nationalize coal, rail and other basic industries and grant welfare reforms in the period following the Second World War. It was acting in response to the threat of revolution posed by the sustained upsurge of the working class and the oppressed masses in the then British Empire, South East Asia and postwar Europe in the mid- to late 1940s and the need to restabilize and restructure bankrupt British capitalism. The nationalizations of basic industries were bound up with the need for coordination and planning to better serve the needs of industry.

Conversely, the privatization of the state-owned enterprise was bound up with the decline of the nationally regulated economy. In the case of the railways, to cite but one example, the decline in mass industrial manufacturing, the disappearance of the nationally based coal industry and the transfer of freight to road transport meant that industry no longer had a significant stake in the railways: nationalization had lost its *raison d'etre*. Rail thus became a target for privatization. The subsidies that had once been given to the public monopolies were now to be given to private sector corporations, marking a new stage in the decline of British capitalism and the on-going assault on the working class.

But there was also another significant factor. The erosion of the welfare state and the sale of the nationalized industries not just in Britain but all over the world in the last 30 years have been directly bound up with the absence of any politically conscious movement in the working class. The socialist conceptions that animated large sections of workers in the aftermath of the Russian Revolution came under sustained attack from Stalinism, labour reformism and trade unionism, which together attacked genuine socialism and above all, internationalism. Such strikes as were held to oppose plant closures in the face of global competition, privatization and other neo-liberal policies were always firmly under the control of the labour bureaucracies and limited to narrow economic and above all to national demands.

To take but one example, the opposition to the London Underground PPPs: the Mayor of London, backed by many radical groups, won the mayoral election in 2000 on the basis of his opposition to the privatization of London Underground. But while he opposed the particular form of privatization, his own proposals were not significantly different: to open up London Underground to the City and big business by financing the investment programme with public bonds (something that local authorities were not at that time allowed to do) and outsourcing the work to private contractors. This, he argued, was a more viable model than the PPP. While the Mayor sought to challenge the PPPs via the courts, this avenue was soon exhausted. The rail unions for their part, some of the Mayor's most fervent supporters, despite calling the new boss of Transport *for* London a 'union buster', refused to organize joint action against the PPP even though rail workers had voted three times for joint action. The Rail Maritime and Transport union repeatedly called off strikes and overturned ballots for strike action in an effort to demobilize workers' opposition to the privatization of the Tube. Such one-day strikes as were held in 2001 were the result of unofficial action. While workers saw strike action as a means of opposing privatization, the unions refused to make this their explicit aim and even refused to make any industrial action against the PPP official. So badly publicized was the limited action that, according to the BBC, most people did not even know what it was about, believing it to be in support of pay rises. The Mayor himself vigorously opposed any industrial action and lambasted the strikes. Yet the 'lefts' continued to support him. The PPPs were pushed through in the teeth of widespread popular opposition and critical reports from the board responsible for running the Underground, the National Audit Office and financial and transport analysts, and transferred to private contractors in April 2003.

Such opposition was therefore largely unsuccessful and, at best, succeeded only in postponing such measures. It illustrates that the defence of public services entails the conscious recognition that capitalism, not the increasingly global character of modern society, is the real enemy. Global capitalism – i.e., the subordination of humanity to the profit interests of a few hundred giant transnational corporations – cannot be fought by seeking to return to a historically outmoded system of relatively isolated and unintegrated national economies. The point is not to reject the advances of science and technology or to return to local, small-scale production, but to wrest control of the enormous productive forces created by human labour from the TNCs and nation states, and place them under the common ownership of all humanity, with their development subordinated in a rational and planned way to human needs. This task requires the development of a conscious political movement of the working class, capable of challenging the very basis of the profit system itself. Such a movement must be armed with a scientific socialist perspective and firmly grounded in the strategic experiences of the twentieth century. It implies a very definite social orientation and programme – that of revolutionary scientific socialism as opposed to the dominant, politically liberal programme of national reformism – which is why the traditional organizations of the working class and intellectuals have been unable to mount any effective opposition to or critique of the neo-liberal agenda.

In conclusion, this chapter has shown that the PFI appraisal methodology is conceptually flawed. The case for PFI rests upon risk transfer, which is impossible to quantify meaningfully and even more difficult to achieve in practice. Although the private partners, particularly the financiers, are well rewarded, it is impossible to see, owing to the lack of information, the extent to which these rewards are commensurate with the risk. When projects fail, which is not infrequently, risk has been transferred back to the public sector and tax payers and/or on to the public as users, a travesty of risk transfer. Projects have had to be bailed out and/or renegotiated. Far from being a form of British exceptionalism, such experiences mirror those of countries that have implemented concessions and design build finance and operate contracts for infrastructure projects. Yet, governments have shown no sign of abandoning the policy. The high cost has become ever more apparent, creating affordability problems for either the commissioning agency or the public sector as a whole, users and the public at large.

Thus, an explanation for the turn to such a policy that displays such a disparity between promises and the outcomes not just in Britain but internationally must be sought in broader economic and political processes. While this chapter has located the policy within the development of the neo-liberal agenda, it explains its origins not just as a policy shift but as the corporate response to fundamental changes within the workings of the economic system itself and the crisis of the profit system as a whole. The rise of the New Right was further facilitated by the failure of the old labour organizations to launch a political offensive against it, because of its opposition to socialism and internationalism. While many of these neo-liberal policies will unravel at public expense due to their own contradictions, the development of socially progressive policies entails the reorganization of society and the building of an international party to do so.

Notes

* The author gratefully acknowledges research support by the University of Manchester's Research Centre for Socio-Cultural Change (CRESC) which generously supported her in 2006 while working on this project.
1. http://www.partnershipsuk.org.uk/AboutPUK/PUKMission.asp (accessed 28/08/07).
2. Personal communication in response to a request for further information from the author of this chapter.

References

Agile Construction Initiative (1999) *Benchmarking Stage Two Study*, Agile, London.
Appleby, J., and Coote, A. (2002) *Five Year Health Check: A Review of Health Policy 1997–2002*, The Kings Fund, London.
Armstrong, P., Glyn, A. and Harrison J. (1984) *Capitalism since World War II: The Making and Breakup of the Great Boom*, Fontana Paperbacks, London.
Arthur Andersen and Enterprise LSE (2000) *Value for Money Drivers in the Private Finance Initiative*, Treasury Task Force, London.

Audit Commission (2003) *PFI in Schools: The Quality and Cost of Buildings and Services Provided by Early Private Finance Initiative Schemes*, Audit Commission, London.

Boardman, A.E., and Vining, A.R. (1989) 'Ownership and Performance in Competitive Environments: A Comparison of the Performance of Private, Mixed and State-owned Enterprises', *Journal of Law and Economics*, 32(1), April: 1–33.

Boardman, A.E., Poschmann, F., and Vining, A. (2005) 'North American infrastructure P3s: examples and lessons learned', in Hodge, G., and Greve, C., (eds) *The Challenge of Public–Private Partnerships: Learning from International Experience*, Edward Elgar Publishing, Cheltenham, UK.

Centre for Public Services (2002) *Privatising Justice: The Impact of the Private Finance Initiative in the Criminal Justice System*, Report, Sheffield.

Chief Inspector of Prisons for England and Wales (2000) *HM Prison Altcourse*, Report of a Full Inspection 1–10 November 1999, Home Office, London.

Craig, D. (2006) *Plundering the Public Sector*, Constable, London.

Crain, W.M., and Zardkoohi, A. (1978) 'A Test of the Property Rights Theory of the Firm: Water Utilities in the United States', *Journal of Law and Economics*, 21(2): 395–408.

DLA Group (2004) European PPP Report 2004, DLA London.

Edwards, P. and Shaoul, J. (2003) 'Partnerships: For Better, For Worse?' *Accounting, Auditing and Accountability Journal*, 16(3): 397–421.

Edwards, P., Shaoul, J., Stafford, A., and Arblaster, L. (2004) *Evaluating the Operation of PFI in Roads and Hospitals*, Association of Chartered Certified Accountants, Research Report no. 84, London.

Ernst, J. (1994) *Whose Utility? The Social Impact of Public Utility Privatization and Regulation in Britain*, Open University Press, Buckingham, Philadelphia.

Estache, A., and Serebrisky, T. (2004) *Where Do We Stand on Transport Infrastructure Deregulation and Public Private Partnership?* World Bank Policy Research Working Paper 3274, World Bank, Washington, D.C.

Farrell, S. (1997) *Financing European Transport Infrastructure: Policies and Practice in Western Europe*, Macmillan, Basingstoke.

Flyvbjerg, B., Bruzelius, N., and Rothengatter, W. (2003) *Megaprojects and Risks: An Anatomy of Ambition*, Cambridge University Press, Cambridge.

Froud, J. (2003) 'The private finance initiative: risk, uncertainty and the state', *Accounting, Organisations and Society*, Vol. 28: 567–89.

Froud, J., and Shaoul, J. (2001) 'Appraising and Evaluating PFI for NHS Hospitals', *Financial Accountability and Management*, 17(3), August: 247–70.

Gaffney, D., and Pollock, A. (1999a) *Downsizing for the 21st Century*, a report to Unison Northern Region on the North Durham Acute Hospitals PFI scheme, School of Public Policy, University College London.

Gaffney, D., and Pollock, A. (1999b) 'Pump-priming the PFI: why are privately financed hospitals schemes being financed?', *Public Money and Management*, January–March: 55–62.

Gaffney, D., Pollock, A., Price, D. and Shaoul, J. (1999a) 'NHS capital expenditure and the private finance initiative – expansion or contraction?', *British Medical Journal*, Vol. 319: 48–50.

Gaffney, D., Pollock, A., Price, D., and Shaoul, J. (1999b) 'PFI in the NHS – is there an economic case?', *British Medical Journal*, Vol. 319: 116–99.

Gaffney, D., Pollock, A., Price, D., and Shaoul, J. (1999c) 'The politics of PFI and the "New" NHS', *British Medical Journal*, Vol. 319.

Gosling, P. (2004) 'Public out, private in: why private sector recruits are filling key senior public sector positions', *Public Eye*, December.

Greenaway, J., Salter, B., and Hart, S. (2004), 'The Evolution of a Meta-Policy: The Case of a Private Finance Initiative and the Health Sector', *British Journal of Politics and International Relations*, 6(4): 507–26.

Heald, D. (1997) 'Privately Financed Capital in Public Services', *The Manchester School*, 65(5): 568–98.

Heald, D. and Steel, D. (1986) 'Privatizing Public Enterprises: An Analysis of the Government's Case', in Kay, J., Mayer, C., and Thompson, D. (eds), *Privatisation and Regulation: The UK Experience*, Clarendon Press, London.

Health Select Committee (2002) *The Role of the Private Sector in the NHS*, HC 308, Session 2001–02, The Stationery Office, London.

Health Select Committee (2000) *Public Expenditure on Health and Personal Social Services 2000, Memorandum received from Department of Health containing replies to a written question from the Committee*, HC 882 Session 1999–2000, The Stationery Office, London.

Hodge, G. (2005) 'Public Private Partnerships: The Australasian Experience with Physical Infrastructure', in Hodge, G., and Greve, C., (eds) *The Challenge of Public–Private Partnerships: Learning From International Experience*, Edward Elgar Publishing, Cheltenham.

Institute of Public Policy Research (2001) *Building Better Partnerships*, Final Report of the Commission on Public Private Partnerships, IPPR, London.

King, P. (1975) 'Is the Emphasis of Capital Budgeting Theory Misplaced?' *Journal of Business, Finance and Accounting*, 2(1): 69–82.

Luxemburg, R. (1988) *Reform or Revolution?* Pathfinder Press, New York.

Mayston, D. (1999) 'The Private Finance Initiative in the National Health Service: An Unhealthy Development in New Public Management', *Financial Accountability and Management*, 15(3&4): 249–74.

Millward, R. (1982) 'The Comparative Performance of Public and Private Enterprise', in Lord Roll (ed.), *The Mixed Economy*, Macmillan, London, 1982.

Millward, R. (1999) 'State Enterprise in Britain in the Twentieth Century', in Amatori, F., (ed.) *The Rise and fall of State Owned Enterprises in the Western World*, Cambridge University Press, Cambridge.

Molnar, E. (2003) *Trends In Transport Investment Funding: Past Present And Future*, UNESCO and CEMT/CS/12.

Mott MacDonald (2002) *Review of Large Scale Procurement*, Mott MacDonald, London.

National Audit Office (1997) *The Skye Bridge*, Report of Comptroller and Auditor General, HC 5, Session 1997–98, The Stationery Office, London.

National Audit Office (1998) *The Private Finance Initiative: The First Four Design, Build, Finance and Operate Roads Contracts*, Report of Comptroller and Auditor General, HC 476, Session 1997–98, The Stationery Office, London.

National Audit Office (1999a) *The PFI Contract for the New Dartford and Gravesham Hospital*, Report of Comptroller and Auditor General, HC 423, Session 1998–99, The Stationery Office, London.

National Audit Office (1999b) *The Passport Delays of Summer 1999*, Report of Comptroller and Auditor General, HC 812, Session 1998–99, The Stationery Office, London.

National Audit Office (2000a) *The Financial Analysis for the London Underground Public Private Partnership*, Report of Comptroller and Auditor General, HC 54, Session 2000–01, The Stationery Office, London.

National Audit Office (2000b) *The Refinancing of the Fazakerley PFI Prison Contract*, Report of Comptroller and Auditor General, HC 584, Session 1999–2000, The Stationery Office, London.

National Audit Office (2001) *Modernising Construction*, Report of Comptroller and Auditor General, HC, Session 200, The Stationery Office, London.

National Audit Office (2002a) *The Public Private Partnership for National Air Traffic Services Ltd*, Report by the Comptroller and Auditor General, HC 1096 Session 2001–02, The Stationery Office, London.

National Audit Office (2002b) *PFI Refinancing Update*, Report of Comptroller and Auditor General, HC 1288, Session 2001–02, The Stationery Office, London.

National Audit Office (2003a) *PFI: Construction Performance*, Report of Comptroller and Auditor General, HC 371, Session 2002–03, The Stationery Office, London.

National Audit Office (2003b) *The Operational Performance of Prisons*, Report of Comptroller and Auditor General, HC 700, Session 2002–03, The Stationery Office, London.

National Audit Office (2005) *The Refinancing of the Norfolk and Norwich PFI Hospital: How the Deal Can Be Viewed in the Light of Refinancing*, Report by the Comptroller and Auditor General, HC 78, Session 20005–06, The Stationery Office, London.

National Audit Office (2006a) *The Modernisation of the West Coast Main Line*, Report by the Comptroller and Auditor General, HC 22, Session 2006–07, The Stationery Office, London.

National Audit Office (2006b) *Update on PFI Debt Refinancing and the PFI Equity Market*, Report by the Comptroller and Auditor General, HC 1040, Session 2006–07, The Stationery Office, London.

National Audit Office (2007) *Improving the PFI Tendering Process*, Report by the Comptroller and Auditor General, HC 149, Session 2006–07, The Stationery Office, London.

Pollock, A., Dunnigan, M., Gaffney, D., Price, D., and Shaoul, J. (1999) 'Planning the "New" NHS: Downsizing for the 21st Century', *British Medical Journal*, Vol. 319.

Pollock, A., Price, D., and Dunnigan, M. (2000) *Deficits before Patients: A Report on the Worcestershire Royal Infirmary PFI and Worcestershire Hospital Configuration*, School of Public Policy, University College London.

Pollock, A., Price, D., and Player, S. (2007) 'An Examination of the UK Treasury's Evidence Base for Cost and time Overrun Data in UK Value for Money Policy and Appraisal', *Public Money and Management*, 27(2): 127–34.

Pollock, A., Shaoul, J. and Vickers, N. (2002) 'Private Finance and "Value for Money" in NHS Hospitals: A Policy in Search of a Rationale?' *British Medical Journal*, 324: 1205–8.

Pollock, A., Shaoul, J., Rowland, D., and Player, S. (2001) *Public Services and the Private Sector – A Response to the IPPR*, Catalyst Working Paper, Catalyst, London.

Pryke, R. (1982) 'The Comparative Performance of Public and Private Sector Enterprise', *Fiscal Studies*, 3(2): 68–81.
Public Accounts Committee (2003) *Delivering Better Value for Money from the Private Finance Initiative*, HC 764, Session 2002–03.
PWC (2001) *Public Private Partnerships: A Clearer View*, PWC, London.
PWC (2005) *Queen Elizabeth Hospital NHS Trust*, public interest report, PWC, London.
Sampson, A. (2005) *Who Rules This Place? The Anatomy of Britain in the 21st Century*, John Murray (Publishers), London.
Shaoul, J. (2005) 'A Critical Financial Appraisal of the Private Finance Initiative: Selecting a Financing Method or Reallocating Wealth?', *Critical Perspectives on Accounting*, 16: 441–71.
Shaoul, J. (2003) 'Financial Analysis of the National Air Traffic Services Public Private Partnership', *Public Money and Management*, 23(3):185–94.
Shaoul, J., Stafford, A, and Stapleton, P. (2006) 'Highway Robbery? A Financial Evaluation of Design Build Finance and Operate in UK Roads', *Transport Reviews*, 26(3): 257–74.
Shaoul, J., Stafford, A. and Stapleton P. (2007a) 'Evidence Based Policies and the Meaning of Success: The Case of a Road Built Under Design Build Finance and Operate', *Evidence and Policy*, 3(2): 159–80.
Shaoul, J., Stafford, A. and Stapleton, P. (2007b) 'Private control over public policy: financial advisors and the private finance initiative', *Policy and Politics*, 35(3): 479–96.
Shaoul, J., Stafford, A. and Stapleton, P. (2008) 'The cost of using private finance to build, finance and operate the first 12 NHS hospitals in England', *Public Money and Management*, April: 101–8.
Spann, R.M., (1977) 'Public v. Private Provision of Government Services', in (ed.) Borcherding, T.E., *Budgets and Bureaucrats – The Source of Government Growth*, Duke University Press, North Carolina.
Stambrook, D. (2005) *Successful Examples of Public Private Partnerships and Private Sector Involvement in Transport Infrastructure Development*, Final Report under contract with OECD/ECMT Transport Research Centre, Contract # CEM JA00028491, Virtuosity Consulting, Ottawa, Canada.
Standard and Poor's (2003) *Public Finance/Infrastructure Finance: Credit Survey of the UK Private Finance Initiative and Public–Private Partnerships*, Standard and Poor's, London.
Stretton, H., and Orchard, L. (1994). *Public Goods, Public Enterprise and Public Choice: The Theoretical Foundations of the Contemporary Attack on Government*, St. Martin's Press, New York.
Taylor, G. (2005) 'Major roads works ahead: 10 years of the UK private finance initiative roads programme', in *Public Private Partnerships: Global Credit Survey 2005*, Standard and Poor's, London.
Timms, S. (2001) *Public Private Partnerships, Private Finance Initiative*, Keynote Address by the Financial Secretary to the Treasury to the Global Summit, Cape Town, 6 December.
Treasury (1991) *Economic Appraisal in Central government. A Technical Guide for Government Departments.* HM Treasury, London.
Treasury (1997) *Step By Step Guide to the PFI Procurement Process*', HM Treasury, London.

Treasury (2003a) *PFI Meeting the Investment Challenge*, HM Treasury, London.
Treasury (2003b) *The Revised Green Book*, HM Treasury, London.
Treasury (2007) *PFI Statistics: Signed Projects List*, http://www.hm-treasury.gov.uk/media/3/6/pfi_signeddeals_110707.xls (accessed 31 August, 2007).
Worthington, J. (2002) *2020 Vision: Our Future Healthcare Environments*, Report of the Building Futures Group, The Stationery Office, London.

2
Private Finance Initiative and Public Private Partnerships: The Key Issues

Malcolm Sawyer

University of Leeds

Abstract

This chapter considers the scale of the Private Finance Initiative (PFI) in the UK, followed by an examination of the argument that PFI is a means of providing additional public investment. The ways in which PFI and assets are treated in public sector accounts, creates biases in decision making. The cost of finance is higher under PFI arrangements than under 'conventional' public investment. There is critical evaluation of two arguments to the effect that although the cost of finance is higher there are compensating advantages for PFI. First, that risk is transferred from the public sector to the private sector, and that the higher price of finance is the 'price' of that risk transfer with private contractor better equipped to deal with risk. The second claimed advantage is that a PFI project is more likely to be delivered on time and on budget, and to a lower budget, as compared with 'conventional' public investment. The chapter is completed by discussion of some contractual issues surrounding PFI. The overall conclusion of the chapter is that PFI is one element of privatization that has been a high-cost way of undertaking public investment.

Keywords: Private Finance Initiative, Privatization

Journal of Economic Literature classification: L33

1. Introduction

In many countries there have been moves, under a variety of labels, to involve the private sector in the financing, construction and operation of capital assets which are used for the provision of public services. In the UK the general title used is public private partnerships (PPP hereafter) within which there is the Private Finance Initiative (PFI hereafter); elsewhere terms such as BOT (Build, Operate, Transfer) are used. The purpose of this chapter is to undertake a critical evaluation of a range of the key issues connected with PFI in the UK, treated as an example of the general trends referred to above. The chapter views PFI as an element of privatization, and the particular focus of the chapter is whether PFI provides additional investment in a cost effective manner.

A public private partnership has been defined as 'a risk-sharing relationship based on a shared aspiration between the public sector and one or more partners from the private and/or voluntary sectors to deliver a publicly agreed outcome and/or public service' (Grimsey and Lewis, 2004, p. xiv). The specific features of the UK PFI programme is that it involves assets and facilities to be used in the provision of public services being constructed and financed by a private sector company with the leasing of those assets and facilities to the public sector along with the provision of services related with the facilities.

The PFI (and PPP more generally) is a particular example of ideas of injecting private sector involvement into the provision of public services. The term 'conventional' or 'traditional' public investment is used here to describe a situation where a capital project for use in the public sector (e.g. a school) is constructed by the private sector and paid for by government out of tax revenue or its own borrowing, and the corresponding capital asset (e.g. school) is then owned and operated by the public sector. The difference between that 'conventional' public investment and PFI can be seen to involve three elements.

> First, with PFI, the building is technically not owned by the public sector – although who has the 'asset' of the building is a major disagreement in PFI.... Second, the design of this building, along with accompanying services, is the responsibility of the private sector. The public sector should not be actively involved in the specification – all it specifies is the outputs it requires it terms of services. Third, the public sector is locked into a long-term relationship, specified as best as possible through a legal contract, with a private sector supplier who might have different values and interests. A genuine concern to many is that this private sector supplier, with its profit emphasis and necessity to give priority to its shareholders, may or may not share the same public service values that might be the case if provision were exclusively made by those in the employment of the public sector. (Broadbent and Laughlin, 2003, p. 336)

Thus PFI involves the private ownership of assets which are used in the public sector and the private provision of some services (such as maintenance and cleaning) associated with the assets.

The Private Finance Initiative can be viewed as a significant aspect of privatization in the sense that assets which under 'conventional' public investment have been owned and operated by the public sector are under PFI owned and operated by the private sector. More broadly it can be viewed as closely linked with the so-called New Public Management (NPM).

> NPM describes a set of practices related to the introduction of approaches that are more akin to those of the private sector.... Initially these were related to the introduction of quasi-markets and competition. At one extreme there was the complete privatisation of a number of previously run services. For those that could not be privatised, often for political reasons, other competitive schemes were developed (for example the quasi-markets in the National Health Service in the UK). The key philosophy was that competition would bring efficiency in the sense that the market would ensure that the weakest organisations would fail as they could not attract the resources required to ensure their survival. (Broadbent and Laughlin, 2005a, p. 78)

This can be seen as reflected in the way in which the present UK government views PFI. It is seen as

> a small but important part of the government's strategy for delivering high quality public services.
>
> In assessing where PFI is appropriate, the government's approach is based on its commitment to efficiency, equity and accountability and on the Prime Minister's principles of public sector reform. PFI is only used where it can meet these requirements and deliver clear value for money without sacrificing the terms and conditions of staff.
>
> Where these conditions are met PFI delivers a number of important benefits. By requiring the private sector to put its own capital at risk and to deliver clear levels of service to the public over the long term, PFI helps to deliver high quality public services and ensure that public assets are delivered on time and to budget. (UK Treasury web site, July 2007)

This statement is rather less bullish with regard to claims on the range of benefits of PFI than some earlier ones, and specifically does not make claims that investment through PFI would not have otherwise taken place (as evidenced in the discussion in section 3).

The statement does though point to a number of crucial elements in PFI, namely the 'value for money' argument and the private sector puts 'its own capital at risk'. The 'value for money' is examined below in two respects, namely whether PFI involves more expensive finance (as compared with government borrowing on its own account and purchasing the corresponding investment) and whether the costs of the construction and the subsequent 'running costs' are less than under the 'traditional' arrangements. This can also be related to the question as to whether projects under a PFI are more likely to be 'delivered on time and on budget' as compared with those under 'traditional arrangements'. The idea that PFI transfers risk from the public sector to the private sector has become a particularly important one, which we explore at some length below. The significance of this transfer of risk argument comes from a number of directions but two are particularly important. First, it is now widely acknowledged that the cost of finance under PFI is higher than under 'conventionally' financed public investment, and it is then argued that this higher cost of finance is in effect the price paid for the transfer of risk from public to private. Second, the transfer of risk leads to changes in behaviour by both the public and the private sector. Specifically it is argued that 'the transfer of risk to the private sector provides an incentive for private entities to maximize efficiency' (Grimsey and Lewis, 2004, p. 34). This raises the question as to whether risk is effectively transferred from public to private and whether the process of PFI creates new risks, such as inability of the contractor to fulfil the contract and inflexibilities in the use of buildings provided under PFI.

The process leading to a PFI project involves the comparison between the proposed project under PFI with a Public Sector Comparator (PSC) where the latter is a comparable investment project which would be funded directly by government and the subsequent provision of services (such as maintenance) which would be included in the PFI. The ways in which these comparisons are undertaken are, of course, crucial to decisions on whether PFI is used, and the use of a hypothetical alternative (the PSC) is inevitably problematic. Assumptions have to be made about the differences in construction costs under the two scenarios, for example. In terms of this PFI–PSC comparison we focus below on issues of differences in the cost of finance and differences in the costs of construction. The construction of the PSC is surrounded by problems, especially when

> the PFI is too often seen as the only option. To justify the PFI option, departments have relied too heavily on public sector comparators. These have often been used incorrectly as a pass or fail test; have been

given a spurious precision which is not justified by the uncertainties involved in their calculation; or have been manipulated to get the desired result. Before the PFI route is chosen departments need to examine all realistic alternatives and make a proper value for money assessment of the available choices. (House of Commons Committee of Public Accounts, 2003a, p. 3)

This chapter begins in the next section by considering the scale of PFI in the UK, which is followed by an examination of the argument that PFI is a means of providing additional public investment. There are a range of issues over the way in which PFI and assets are treated in public sector accounts, and we are interested in whether the ways in which they are treated biases decision making. In the following section the argument is put that the cost of finance is higher under PFI arrangements than under 'conventional' public investment. This leads us into a consideration of two arguments to the effect that although the cost of finance is higher there are compensating advantages for PFI. The first of those is that risk is transferred from the public sector to the private sector, and that the higher price of finance is the 'price' of that risk transfer. Further, the private contractor is better equipped to deal with risk. The second claimed advantage is that (partly as a response to the transfer of risk) the project is more likely to be delivered on time and on budget, and to a lower budget, as compared with 'conventional' public investment. The chapter is completed by discussion of some contractual issues surrounding PFI.

2. Scale of PFI

This section provides some background on the relevant dimensions of the scale of the PFI in the UK. The first refers to the scale of investment in assets used in public sector provision which is financed in this way. Table 2.1 provides figures for PFIs broken down by government department responsible for the commissioning of the PFI projects. It can be readily seen that the PFI is concentrated in the area of transport, defence, health (often hospital construction), and education (school construction and refurbishment).

The totals given at the bottom of Table 2.1 indicate the overall investment arising from the PFI programme in 2006/07 and 2007/08. In relation to overall public expenditure the figures (at around £7 bn in 2006/07) appear relatively small: overall public expenditure is estimated at around £552 bn for 2006/07, and hence PFIs were equivalent to just over 1 per cent of public expenditure. A more appropriate comparator would be other forms of public investment: gross investment in 2006/07

Table 2.1 Private Finance Initiative: estimated capital spending by the private sector (signed deals) (projects in £million)

	2006–07	2007–08
Educations and Skills	759	454
Health	1025	1008
Transport	1435	1110
Communities and Local Government	107	140
Home Office	23	20
Constitutional Affairs	15	2
Defence	557	643
Foreign and Commonwealth Office	5	0
Trade and Industry	1	4
Environment, Food and Rural Affairs	128	249
Culture, Media and Sport	38	12
Work and Pensions	46	55
Scotland	595	618
Wales	0	20
Northern Ireland Executive	98	114
Chancellor's Departments	2	2
Total	**4834**	**4450**

Note: PFI activity in local authority projects is included under the sponsoring central government department.
Source: HM Treasury, *Budget Report* 2007, Table C17.

was estimated at £43.2 bn, and after depreciation and asset sales, net investment undertaken directly by the public sector at £ 25.5 bn. Thus it could be said that over one-fifth of the overall net capital investment contracted by the public sector (that is the £7 bn under PFI and the £25.5 bn directly undertaken) was being undertaken through the PFI.

Under the PFI, the government is contractually committed to lease the project from the private sector company for a specified period (often 25 to 30 years) ahead (and on the other side the private company is contractually obliged to lease the project to the public sector). For the private company this provides a guaranteed future income stream (usually in real terms with price to be paid for lease and services linked to the retail price index). For the government, there is the contractual obligation to make those future payments, providing a valuable asset for the private provider which may be sold on to others.

Table 2.2 provides an indication of the extent of future commitments that flow from the PFI. The figures indicate the future obligations from the contracts in force in March 2007 (at the time of budget report). The expectation would be that as more PFIs come on stream in the

Table 2.2 Estimated payments under PFI contracts – March 2007

	£ billion		£ billion
2006–07	6.9	2019–20	5.8
2007–08	7.3	2020–21	5.9
2008–09	7.8	2021–22	5.5
2009–10	8.2	2022–23	5.4
2010–11	8.5	2023–24	5.4
2011–12	8.6	2024–25	5.4
2012–13	8.7	2025–26	5.2
2013–14	8.8	2026–27	5.0
2014–15	8.8	2027–28	4.8
2015–16	8.8	2028–29	4.5
2016–17	8.9	2029–30	4.2
2017–18	8.2	2030–31	3.8
2018–19	5.8	2031–32	3.4

Source: HM Treasury: *Budget Report* 2007.

future, these expenditure commitments from the PFI programme will rise significantly. It can be seen from Table 2.2 that at present the future obligations amount on an annual basis to around £8–£9 bn (calculated in 2007 prices) for the next decade or more. This compares with a total public expenditure in 2006/07 of £552 bn, and government borrowing of £37 bn.

The cumulative figures in Table 2.2 give a future commitment of £169.6 bn. A (real) discount rate of 3.5 per cent would put present value of those commitments at £120 bn and a 6 per cent rate £98 bn. The 3.5 per cent rate was chosen as the newly recommended rate for project appraisal (HM Treasury, 2003b Annex 6) and 6 per cent as a figure often used in PFI evaluations. By way of comparison, the public sector net debt stood at £500 bn in March 2007 (and £568 bn calculated on the Maastricht treaty definition). The public debt to GDP ratio has tended to decline in the UK over the past decade, but a significant part of that decline can be attributed to the use of PFI in that an equivalent amount of investment undertaken in a 'conventional' manner would have added to the public debt to an extent that would have meant constant or rising public debt to GDP.

These figures provide some indication of the scale of PFI in the UK. There are issues to which we return below on the ways in which the scale of PFI is measured, and specifically the extent to which the activities are on or off the balance sheet. A significant question is the effect the PFI programme has on fiscal policy decisions in that PFI (compared with public investment financed by government at time of construction) shifts expenditure from capital account to current account and from the present into the future.

3. Additional finance and additional investment?

A major argument put forward for PFI and similar arrangements which was much heard in the early days but rather less recently is that PFI provides additional funds and additional investment for the public sector. There has been a widespread perception that constraints on public expenditure and pressures to restrain budget deficits have tended to lead to reductions in capital expenditure rather than in current expenditure. The argument is straightforward – reductions in current expenditure are quickly noticed – they may involve cuts in education or health provision or in social security payments – in a way in which cuts in capital expenditure are not. It is then politically expedient to cut capital rather than current expenditure. The record for the UK (as well as many other countries) on public capital expenditure shows the large drops during the 1970s through the 1990s. The general trend has been described in the following way.

> Government budgets have come under pressure, and with that development have come two responses. One has been a desire to reduce the impact of infrastructure spending on government budgets, both as a means of minimizing government borrowing and as a way of protecting economically necessary but politically dispensable infrastructure expenditure from general budgetary pressures. The other response has seen governments turn to private capital markets for infrastructure funding. (Grimsey and Lewis, 2004, p. 30)

The PFI was then seen as a way of reversing the declines in public investment without an increase in recorded public expenditure or budget deficit, Although that is illusory since while PFI does not add to recorded public expenditure in the years when the investment is taking place it does raise (often over a 30 year period) future public expenditure.

The argument that PFI was a way of providing additional public investment was reflected in statements such as that made by, for example, Paul Boateng, then Financial Secretary to the Treasury, who argued that

> for the future we plan 100 new hospital schemes – including 26 new PFI hospitals to be up and running by mid 2005 – eight already up and

running, 15 more at various stages of construction. An investment in our nation's health that would not have been possible without finance from the private sector. (Boateng, 2001)

In a similar vein, another government minister has argued that

The ability to raise capital from the markets does of course require the prospect of a commercial return – something which is not intrinsically on offer for facilities such as hospitals or schools or prisons where we do not charge the customers. But the Private Finance Initiative, with the concept of paying for a service over 20–30 years, rather than paying for a building in one go, has created a means of providing a commercial return on such projects. Thus we have a vehicle for private capital in projects which were previously the unchallenged domain of public funding. (Falconer, 2001)

The argument that the PFI provides additional finance for the public sector and enables projects to proceed which would not otherwise have been possible is essentially spurious. In whatever way, investment in or for the public sector is undertaken there are resource costs involved in the construction and maintenance of the investment project, and the investment has to be funded whether directly or indirectly. As Sussex (2001) argues,

if we observe that more NHS investment is made now with the PFI than was made before it was introduced, then this is the result simply of a political decision to increase investment. Given the government's current tests of fiscal prudence, there appear to be no current macroeconomic reasons for preferring PFI to Exchequer financing, or for regarding one approach as any more affordable than the other.

Consider what the limitations on the level of public investment actually are. In a fully employed economy, the resources used in one direction are not available for use elsewhere. The limit on public investment is then the availability of resources: the opportunity cost of the investment is that resources have to be drawn away from use elsewhere. In an economy with spare resources of labour and capital, that would not be the case. Resources are available that can be put to use (whether to construct public investment projects or for use elsewhere is a political decision to be made). But the limits are in effect the same whether the investment programme is undertaken through PFI or through some other form.

The limitation on public sector investment may be thought to be ones of finance and funding. An investment project may be financed from general taxation, which would mean that tax revenue is higher (than it would be otherwise) and finance available for private expenditure thereby reduced. When a government decides that capital expenditure is not to be financed through general taxation but through borrowing it faces, in effect, a choice. The public sector can borrow from the private sector and uses the funds obtained to pay for the investment project, or the private sector finances the project directly as through the PFI schemes. Whether directly or indirectly, the public sector is in effect borrowing from the private sector, and the investment funds are obtained through different routes. The public sector repays the full cost of the private sector companies which have constructed the investment projects in annual payments over periods of 20 to 30 years. But it does not provide access to any higher level of funding than would otherwise be the case with public funding. In either case, the public sector faces future obligations – either in the form of future interest payments on the borrowing or in the form of leasing and other charges to the private sector.

4. Accounting issues

The manner in which the expenditures and debts associated with PFI activities are recorded (or not) in the public accounts will matter when decisions made by government are influenced by the state of the public finances. When there is concern over the size of the budget deficit, then the exclusion of PFI expenditure from the current budget deficit may prejudice decision making in favour of PFI (even though PFI adds to future budget deficits). Similarly, a focus on the size of public debt (relative to GDP) also favour PFI when the liabilities to future expenditure arising from PFI are not included in the debt calculations.

The first issue arises from the way in which government capital accounts are presented, namely that liabilities are reported but assets are often largely forgotten. Financial liabilities such as government bonds can be readily valued whereas physical assets (e.g. infrastructure) are more problematic to value. Many of the liabilities have a market value whereas there is not much of a market in second-hand roads and hence the calculation of asset values difficult. It is generally the case that the central focus is on the size of the public debt without taking into account the scale of the public assets. This is clearly reflected in the Stability and Growth Pact (and previously the convergence criteria under the Maastricht Treaty) with the constraint (not, of course, observed in practice) that public debt should not exceed 60 per cent of GDP. Leaving aside the question of why 60 per cent is seen as such a crucial figure, the point here is that a constraint has been put in place, and that the constraint refers to a particular measure of government liabilities. The concern expressed relates to the liability side of the balance sheet with no reference to the asset side. In the context of the public sector, assets do not usually directly generate income for the government, which may explain a focus on liabilities rather than assets. But it does ignore the fact that public sector assets are productive in a more general sense and add to the productive potential of an economy and thereby output and to tax revenue. Assets should be included in any evaluation of the government's balance sheet position, rather than merely to focus on the liability side.

A second issue is that the liabilities of government only include the liabilities incurred in the form of financial assets and do not include commitments to make future payments. Hence it neither includes commitments to make future transfer payments such as pensions, nor (and more significantly here) does it include commitments to future payments under leasing and service agreements (as in PFI). Private sector accounting practice would include leases over three years as a future liability in the capital accounts (balanced by a corresponding asset); public sector accounting does not.

A PFI project involves the creation of both assets and liabilities. But it often leads to the investment project being 'off balance sheet' as far as the public sector is concerned in so far as the assets and liabilities do not appear on the balance sheet of the public sector. Placing projects off balance sheet is likely to influence decisions on investment and the use of PFI. Two sets of accusation have been made in this regard. First, that there may be a drive to get projects 'off balance sheet' in order to limit the apparent size of the government's budget deficit and debt obligations. Second, that some projects may 'disappear' from view and not appear on either the balance sheet of the government or on that of the private company undertaking the PFI project.

The UK government has recently denied that these balance sheet considerations have played any role in the development of the PFI:

> The decision to use PFI is taken on value for money grounds alone, and whether it is on or off balance sheet is not relevant. Almost 60 per cent of PFI projects by value are reported on Departmental balance sheets and fully reflected in the Government's national accounts.

> The Government publishes a complete statement of the costs of PFI facilities, which are fully covered by annual unitary payments, in the Budget document. (HM Treasury 2003a, p. 13)

Similarly,

> the Government only uses PFI where it offers value for money, considered over the long term.... The financial reporting and balance sheet treatment of projects are subsequent and irrelevant to the decision whether to use PFI, but the monitoring and reporting of financial commitments made under PFI is an important part of managing the public finances. (HM Treasury, 2003a, p. 22)

But governments do set targets in terms of debt to GDP and/or budget deficits to GDP ratios, and if these targets have any meaning then the manner in which those ratios are calculated would have some effect on what governments decide to do. In the first years of PFI the borrowing associated with the PFI does not appear on the government's balance sheet, and the size of the budget deficit is lower than it would have been if expenditure on 'conventional' public sector investment had taken place. When there is political concern and pressure over the size of budget deficit, the use of PFI appears a way around the perceived constraints imposed by the size of the budget deficit. But, of course, the PFI does involve future expenditure by government (stretching over 25 to 30 years). In due course, this is likely to place rather greater pressure on the level of public expenditure than would have been the case through 'conventional' public expenditure for two reasons.

First, the adoption of something like a 'golden rule' (that governments can borrow to fund capital expenditure but not, over the course of the business cycle, to fund current expenditure) would mean that there would be limits on current expenditure (in line with taxation) but no specific limits on capital expenditure. The 'golden rule' has been formally adopted by the UK government, and there is considerable discussion over amending the Stability and Growth Pact along similar lines so that budget deficit limits apply to current budget only.

Second, the time profile of expenditure under PFI is rather different to that under the financing of 'conventional' investment. With 'conventional' public investment, the costs (to the government) are incurred at the outset with the costs of construction whereas under PFI the immediate costs are just those of contracting and the leasing costs on PFI are spread over the 30 years or so of the contract. Hence the costs of 'conventional' public investment are much more 'front loaded' than with PFI. The choice of discount rate then becomes of some significance and the higher the discount rate chosen the more likely it is to favour PFI over 'conventional' public investment.

The discounted present value of the two streams of expenditure are generally of similar orders of magnitude, but the total (undiscounted) expenditure under PFI will be greater than under 'conventional' financing. Consider a project which may be financed either through PFI or 'conventional public investment', and where the net present values of the two alternatives are similar. The costs (to the government) of the former will appear later in time than is the case for the latter. The cumulative (real) costs would be lower for the 'conventional public investment' than for the PFI. Even allowing for growth, of GDP, the costs of PFI relative to GDP will be higher than the costs of the 'conventional public investment'. Although PFI reduces pressures on public expenditure in its early years, it places much more pressure in later years. In an era where emphasis of macroeconomic policy making is placed on the current size of the budget deficit, there are incentives to use PFI as it (in comparison with the use of 'conventional' public investment) lowers the reported budget deficit now, albeit at the expense of higher reported budget deficits in the future.

The problems that arise through the ability of companies to shift assets and liabilities off balance sheet have come to the fore in the context of the sub-prime lending crisis. The disquiet over these practices is shown in the following.

> Off-balance sheet accounting has a long and often dishonourable history. Just think of Enron. The Treasury now has a chance to fix one of the most inconsistent and problematic examples of its use, in the way the private finance initiative is accounted for, by using the introduction of International Financial Reporting Standards as a chance to bring all PFI projects on to the public balance sheet.
>
> Britain, and many other countries including Canada, Japan and Australia, use PFI to pay for infrastructure. Instead of paying a contractor to build a school, the government will pay to lease it over 30 years, transferring some management and operating risk to the builder. The question is whether that 30-year lease commitment is a debt, in which case it should be on the balance sheet, or not.
>
> The UK's answer to this question has been inconsistent, to say the least. Most transport PFIs are on the balance sheet. Many hospital PFIs are not. The result is a deep-seated public suspicion that disguising

debt, and not advantages from transferring risk or better private sector management, is the real reason for for using PFI.

The government now has a chance to correct that impression, and allow private finance to show its genuine advantages, when it adopts IFRS in April next year. (*Financial Times*, editorial, 27 July, 2007)

As the *Financial Times* notes, such a re-classification would raise the reported debt to GDP ratio and would require at a minimum that the present government's fiscal rules of 40 per cent debt to GDP ratio would need to be modified (if not modified then it would be broken).[1]

The conclusion can then be drawn that '[o]nce PFI is on the books, the distinction between PFI and conventional capital will disappear. The only reason then for going down the PFI route will be because it transfers risk to the private sector, making the higher cost of borrowing worthwhile' (Timmins, 2007). This leads us to a discussion of the relative costs of finance under PFI compared with conventional public investment, and whether the higher costs of finance under PFI are warranted by the transfer of risk from the public sector to the private sector.

Broadbent and Laughlin (2005b) contains an interesting discussion of the tensions between the UK government and the Accounting Standards Board over the accounting treatment of assets and service commitments arising from PFI projects. But as Heald remarked over a decade ago disquiet and criticisms over the accounting treatment of PFI 'is likely to come from those concerned about 'honest government accounting', a group that is neither numerous nor political weighty' (Heald, 1997, p. 594).

5. Cost of finance

The use of a PFI scheme rather than 'conventional' funding of public investment in effect replaces direct borrowing by government with indirect borrowing, and moves from government being able to borrow at the lowest rates (on government bonds) to borrowing by private companies through more expensive channels, which will be reflected in the charges to government made for the PFI scheme. It is widely accepted that the government can borrow at the lowest rates of interest simply because government bonds are regarded as free of default risk as the government has powers of taxation and to issue money which ensure that the government can always pay. Borrowing by private sector involves some risk as far as lenders are concerned, even if the risk is regarded as rather small.

A key obstacle to the use of private finanance in public projects is that private finance is always more expensive. Therefore, private sector financing must bring with it operational efficiency gains whose prsent value exceeds the present value of additional financing costs. Those advocating the use of private finance contend that these efficiency gains do outweigh the high financing costs. This argument is rarely empirically documented, and is normally asserted to be a consequence of the better incentive structures which prevail in the private sector. Crucially, the driver for efficiency gains is argued by the Treasury to be the transfer of risk to the private sector. (Heald, 1997, p. 574)

Another aspect which points towards the additional costs of finance associated with PFI comes from the refinancing of PFI projects. The risks associated with a proposed PFI project change between the pre-contract period (when there is still uncertainty over the award of the contract as well as its precise terms) and the post-contract period. This can then be reflected in a lower cost of finance (for the PFI contractor) in the post-contract period than in the pre-contract period. The ability to re-finance the PFI contract can then be an additional source of profit for the contractor. A report from the National Audit Office summarizes the argument.

> This figure shows that, once the required service has been brought into operation, the project risks are lower, as the risks associated with commencing service delivery are no longer relevant. This creates opportunities to reduce the annual financing costs, as funders are prepared to offer better terms for projects with lower risks.... Lower annual financing costs improve the returns that can be paid to the private sector shareholders. (National Audit Office, 2002, p. 1)

The realization of the gains to the private company of refinancing lead to PFI deals including arrangements for the sharing of refinancing benefit between the government and the private contractor on a 50–50 basis (see, for example, National Audit Office, 2002, p. 23). But four years later, the expected benefits had not been realized. The Public Accounts Committee (of the House of Commons) found that

> [m]ost of the negotiating of refinancing deals is undertaken by the public sector at a local level where officials often lack commercial awareness. Some of the locally negotiated refinancings have produced very high investor returns and increased risks for the public sector

such as higher termination liabilities and longer contract periods.... To date, proceeds for government from refinancing under the voluntary code amount to only £93 million. This amount is well below the indication provided by the Office of Government Commerce at the PAC hearing in 2003 that proceeds would be in the order of £175–200 million. The shortfall may be a reflection of some investors opting to defer refinancing in favour of realizing gains through selling their shares in the secondary equity markets (House of Commons, Committee of Public Accounts, 2007b, p. 5)

It is beyond dispute that government is able to borrow at a lower rate of interest than the private sector, and hence the cost of finance under 'conventional' funding of public investment will be lower than under PFI. The extent of the difference and how far the additional cost is justified by risk transfer may be matters of debate. In a general sense a PFI can be seen as involving the private company involved in some risk – if, for example, the costs of providing the maintenance of the buildings over the 25 to 30 years of the contract vary with the price paid by government having been agreed (in real terms), then the private company gains or loses in terms of profits as the costs are lower or higher than anticipated. In effect the question can be posed as to whether the higher cost of finance under PFI is compensated by the transfer of risk from public to private sector and the behavioural responses to that risk transfer.

6. Risk sharing

The Treasury has sought to address the argument that PFI finance is more expensive than 'conventional' public investment finance in the following way. It argues that

> there are two common assertions made to justify the claim that the private sector cost of capital exceeds the public sector cost of capital:
> 'Governments can borrow at a risk free rate of interest'.
> This is not the case, there is a risk premium either way, it is just explicit in the price of private capital. Where gilts are used, taxpayers effectively underwrite the associated risk and the price reflects this fact. The taxpayer takes on the contingent liability, and where the risk materializes, they carry the cost as a result. If the taxpayer were to be compensated it would be equivalent to paying the risk premium at the point of raising the capital, making the public and private sectors' cost of capital equivalent.

'The government is better at diversifying the risk than the private sector'.

This assertion is based on the Arrow–Lind theorem, an academic theory which assumes that project returns can be treated as wholly independent of National income. In fact this is rarely the case as public investment is not risk free. (HM Treasury, 2003a, p. 124)

The government is generally regarded as borrowing at a risk free rate in the sense that the default risk is perceived as zero – the government will always be able to pay its debts through taxation or through issue of money. But the operation of the assets acquired (e.g. the school, the hospital) is subject to risk in the sense that the costs may vary for reasons outside the control of the government. However, as argued below, the government in effect pools such risk across a number of projects.

The basic argument

The general idea is that any investment (whether in the public sector or in the private sector) carries risks and the notion that appropriate allowance for risk and risk bearing should be made have become central to the debates over PFI. The cost of finance under different regimes (e.g. PFI, 'conventional' public sector investment) may reflect differences in the bearing of risk. In effect it is argued that under 'conventional' public sector investment, the government can borrow at a relatively low rate of interest, which is perceived to be risk free (as far as the lenders are concerned), but the government bears the risks associated with the operation of the public sector investment. Under the PFI the company concerned borrows at a higher rate of interest, which is reflected in the price it charges the government, but the company bears the risks associated with the operation and maintenance of the investment project. An early expression of the argument is:

An important assumption underlying the PFI is that private sector management will reduce variability risk (e.g. reduce capital overspends or operational underperformance) reflecting the proposition that risks ought to be borne by those best able to control them. It is believed that the private sector will exercise more resilience to cost overruns because these contractors know that the private sector firm is itself vulnerable to bankruptcy and lacks government's access to tax revenues. (Heald, 1997, p. 575)

Although government borrowing is from the lenders' perspective one involving virtually no default risk, and the rate of interest at which the government can borrow reflects that, it would still be the case that the project for which the government is seeking funds bears some risk. A school may be built but there are risks associated with the running and maintenance of the school – the roof may leak, a wall may collapse, etc. Under PFI these risks are transferred to the private contractor (although if there is some catastrophic event that threatens the whole project, the contractor is likely to be bailed out by the government). But by parcelling out the risk the amount of risk to be borne is in effect increased. The government operating many schools is in effect pooling the risk.

The approach of the UK government is that '[t]he appropriate sharing of risks is the key to ensuring value for money benefits in PFI projects are realized', and that risks involved in design and construction should be 'borne by the party who is best placed to manage them'. The intention is that

> [w]here risks are transferred, it is to create the correct disciplines and incentives on the private sector to achieve a better outcome.... Successful PFI projects should therefore achieve an optimal apportionment of risk between the public and private sectors. This will not mean that all types of risks should be transferred to the private sector. Indeed, there are certain risks that are best managed by the Government; to seek to transfer these risks would not offer value for money for the public sector. (HM Treasury, 2003a, p. 35)

There are though a range of issues that arise. Even if it is accepted that risk can be shifted and more effectively borne by the private sector, there are questions as to the pricing of risk and how much the public sector has to pay the private sector to bear the risk, and whether the risk is effectively transferred.

The pricing of risk

The way in which risk is priced can be decisive in decisions on PFI. Yet there is little firm basis for the pricing of risk. This is illustrated in the following report by the House of Common Committee on Public Accounts on a PFI project undertaken for the Ministry of Defence (MOD). They concluded that:

> The inherent uncertainties in a financial comparison of this kind are highlighted by this deal. The MOD's central estimate of the PSC cost of £746.2 million included an adjustment for risk of £102.9 million or 17.2 %.... It is right to take differences in risk into account in

these comparisons since it is not possible to predict conventional procurement costs with absolute accuracy. The outcome of the financial comparison was, however, very sensitive to the costing of risk.... For example, a change in the risk adjustment for capital expenditure from MOD's assumption of 29.5% to 29% reduces the PSC by £1 million. MOD's own calculations showed that, depending on the assumptions that were made for all the various risks, the costs of conventional procurement could fall anywhere within the range of £690 million to £807 million.... MOD considered that its risk assessments were not excessive. It noted that its capital expenditure risk adjustment of 29.5 % was lower than the risk factor that Treasury guidance at the time suggested should be applied to construction costs in conventionally procured projects. A more recent student of public building projects commissioned by the Treasury estimated a range for construction cost overruns of between 2 and 24% in standard building projects and between 4 and 51% in those involving non standard buildings. The extent of these ranges highlights how difficult it is to predict with certainty what the costs for a conventionally procured project might have been. (House of Commons Committee of Public Accounts, 2003c, pp. 7–8)

In a further report relating to a PFI undertaken by GCHQ they found that:

The Comparator included £151 million for additional risk as a measure of the average cost overrun of 24% in public sector managed projects. This figure was the percentage given by the Treasury, who said that it was at the bottom of the range and it arose from a study carried out by a firm of consulting engineers. This risk allowance in itself more than accounted for the difference between the PFI bid and the Comparator and had done so in other projects as well.

Other PFI deals had used different, lower risk addition percentages; and with a range of other adjustments available from the use of different discount rates (and the way service costs were spread), it seemed to us that the Public Sector Comparator figures could be used to demonstrate any result required. Such uncertain figures risked clouding the issue of value for money and could cloak a predisposition to go in for PFI. (House of Commons Committee of Public Accounts, 2004)

Others have similarly commented that

in all schemes [considered] risk transfer is the critical element in proving the value for money case. There is considerable variation between schemes in the absolute and relative value of risk transferred. What is striking, however, is that in all cases risk transfer almost equals the amount required to bridge the gap between the public sector comparator and the PFI. This suggests that the function of risk transfer is to disguise the true costs of PFI and to close the difference between private finance and the much lower costs of conventional public procurement and private finance. (Pollock *et al.*, 2002, p. 1208)

Risk is generally viewed as related to the variance of possible outcomes. In the example given above, risk was clearly being measured in terms of the expected costs from delays, etc., and is not related to variability and the costs of variability. Risk is essentially an *ex ante* concept and there are clear difficulties in assessing what the risk may be, and in assessing whether the prior estimates of risk arose in practice (and with PFI only with a 30-year delay). Thus, it can be argued that

risk transfer requires the ability to quantify the probability of things going wrong. There is no standard method for identifying and measuring the values of risk, and the government has not published the methods it uses. The business cases we examined do not reveal how the risks were identified and costed. Our findings are supported by a Treasury commissioned report which found that in over two thirds of the business cases for hospital PFI schemes the risk could not be identified. In the other cases risk transfer was largely attributed to construction cost risks, which would be dealt with by penalty clauses under traditional procurement contracts. (Pollock *et al.*, 2002, pp. 1208)

Let us note that projects differ in the extent to which they are similar to previous projects and the extent to which they can be considered one-off. When a project (e.g. building a school) appears to be similar to previous ones, then experience from previous projects may enable the contract to be specified in terms of a fixed price. The project may involve elements that are new, but with a fixed price the firm undertaking the project bears the risk if costs outrun those expected (and reaps the gains when costs are lower than expected). But when the project confronts the unknown (e.g. the construction of the Channel Tunnel, the development of Concorde), and where the risks may be incalculable, then it is unlikely that a firm would be willing to sign a fixed price contract: simply by doing so, a firm would be putting its existence on the line.

There are clearly issues of risk bearing and of incentives involved here, which may be irresolvable. A fixed price contract pressures the contractor to be cost efficient, but that may be at the expense of the overall quality of the project. But the initial price may be set higher than it would be otherwise to compensate for the risks involved. The contractor undertaking the project cannot know what unforeseen circumstances (positive or negative) may arise. A cost-plus contract means that the risk is retained by the public sector and there are not incentives to minimize costs (or to complete the contract on time). Such a contract can however allow for unforeseen circumstances.

The introduction of PFI does not change the basic problem of how a contract can be written between government and private contractors that both leads to cost efficiency and deals with the inherent risks. A contractor may be awarded a fixed price contract whether that price is paid 'up front' (as a capital sum) or is paid over many years (as a fee for leasing and services). Conversely, a contractor may be awarded a form of cost-plus contractor in either case. The incentives for a contrast to be completed on time and on budget can be secured by the writing and monitoring of appropriate contracts for the construction project with penalties imposed for late delay, etc.

Is 'risk' effectively transferred?

One way in which risk is transferred from the public sector to the private sector under PFI is that the payments under the contract are assured and the contractor accepts the risks associated with the provision of the services under the contract which include variations in costs and impact on profits. Under a 'conventional' public sector investment, the provision of services would be undertaken directly by the public sector which thereby incurs the risks associated with fluctuating costs. The effective transfer of risk would, of course, mean that in the event of a major difficulty threatening the profitability of the PFI project there would be no assistance forthcoming from the government but the PFI contractor would have to bear the costs. There are, not surprisingly, problems arising here from the 'too big to fail' syndrome. An example of this is given in the following:

> The Passport Agency PFI provides an example of the political realities of risk transfer in the context of a high profile, essential service. The fact that compensation was waived and the allocation of the costs of failure negotiable suggests that risk transfer was not after all secured by

the contract, or not to the value contractually specified and in respect of which the risk premium was payment. (UNISON, 2004, p. 33)

A further example is given by the case of the National Air Traffic System:

> This privately-owned service now depends on a government life-support machine, because it cannot be allowed to fail ... [T]he case, precisely because it was a failure, is useful because it raises issues about the rationale, appraisal and risks of the Government's partnerships policy in the context of services that cannot be allowed to fail.... The Public Accounts Committee has twice drawn attention to the paucity of data on the relationship between risk and the cost of private finance.... The expectation that changes in risk transfer are accompanied by changes in the premiums paid to private financiers and adjustments to annual payments has not been tested. (UNISON, 2004, p. 37)

The House of Commons Committee of Public Accounts has noted that:

> Departments are too willing to bail out PFI contractors who get into trouble. Contractors should expect to lose out when things go wrong just as they expect to be rewarded when projects are successful. Departments must ensure that PFI contracts safeguard the taxpayer's position in circumstances where the contractor is no longer able to deliver what is required under the contract. Departments should consider in advance how they will eventually exit from deals should this prove necessary and draw up contingency plans accordingly. When projects run into difficulties prompt action is necessary to prevent costs rising further. They taxpayer must not be expected to pick up the tab whenever a deal goes wrong. (House of Commons Committee of Public Accounts, 2003a, p. 4)

Edwards and Shaoul conclude that

> our analysis has shown that the concept of risk transfer that lies at the heart of the rationale for partnerships is problematic, regardless of whether the project is 'successful' or not. If the project is successful, then the public agency may pay more than under conventional procurement: if it is unsuccessful then the risks and costs are dispersed in unexpected ways. Hence public accountability is obscured.... our analysis shows that, although a project fails to transfer risk and

deliver value for money in the way that the public agency anticipated, the possibility of enforcing the arrangements and/or dissolving the partnership is in practice severely circumscribed for both legal and operational reasons. (Edwards and Shaoul, 2002, p. 418)

Is there a net change in the amount of risk?

The arguments on risk transfer are based on the notion of shifting risk from the public sector to the private sector, and the postulated superior ability of the private sector to deal with that risk. The argument here is that the amount of risk may in a significant sense be increased by moves towards PFI. There are two aspects to the argument. First, there is in effect de-pooling of risk under PFI in that individual firms now bear risk that would otherwise be collectively borne by the public sector. Second, PFI introduces inflexibilities that make it more difficult to cope and respond to risk.

First, when risk is considered in a probabilistic manner and when there is no significant (differential) impact of risk on behaviour, then the public sector benefits from 'the law of large numbers' in terms of risk. Consider the case where a range of similar projects is undertaken, each of which has a stream of costs subject to random factors, and the variance of the stream of costs of an individual project is assessed at σ^2, and this is treated as a measure of risk. The variance of N projects is then σ^2/N. In so far as the random factors are correlated across projects (arising from say common weather conditions or economic circumstances), then the reduction in the variance as N rises will be less than indicated. Thus having the projects undertaken on an individual basis and with the risk of individual projects separately priced, then the degree of risk is then greater for the separate schemes than for the pooled arrangements.

The second is that as far as the public sector is concerned there is a loss of flexibility under the PFI arrangements. It is well known that any contract drawn up for any significant period of time suffers from issues of flexibility in so far as the contract cannot possibly specify reactions to all changes in circumstance. Seeking to do so would entail extremely long contracts, and may not be possible in a world with some degree of uncertainty. The PFI contracts are typically for 25 to 30 years, and specify the services to be rendered over that period. The 'conventional' public sector alternative also includes degrees of inflexibility: once a school is built, its use cannot be readily changed, etc. But it is clear that there is some flexibility: demographic changes may render the school surplus to requirements, its use may be changed and the associated maintenance arrangements, etc. changed. Under a PFI compensation to the contractor would be required, etc. A recent expression of this comes from the NHS Confederation when they write that

> While the healthcare environment is changing rapidly, PFI, with its 30–35-year contractual time horizon, is acknowledged to be inflexible. Once built, it is difficult and expensive to modify any hospital, but the process is more complicated and expensive for PFI projects.... Problems arise for a PFI project when there are unanticipated developments after the contract has been agreed which require changes in the size or design of the new hospital facilities. Most large PFI hospital projects completed so far have had to adapt their planned use or increase in size during procurement or after completion. (NHS Confederation, 2004, p. 7)

There is a sense in which a PFI scheme increases risk for the public sector in that the public sector is committed to pay for a specified project over say 30 years even if the project is no longer of use to the public sector. The PFI approach introduces a significant element of inflexibility, namely that the services related to the capital project are contracted for a period of many years ahead (often 25 or 30 years). If say a school is constructed, there is some element of inflexibility in that there may be few alternative uses for a school building: in other words the construction of a school represents a strong element of 'sunk costs'. But the buildings and the land on which it is built may have some alternative uses. However when the school is constructed under the PFI, then there will a leasing arrangement on the school for 25 or 30 years and often maintenance and other contractual arrangements. Demographic changes over a quarter of a century will bring a changing pattern in the requirements for school places, and yet those schools constructed under PFI will be contracted to continue to provide services whether or not they are wanted. Hence,

> taking on a 30 year contract for services is an additional risk for [NHS] trusts. If the demand for hospital services is reduced for any reason, the NHS trust is still tied into an agreement for maintenance, facilities, and management services over and above the cost of building the hospital. This would not be the case if the hospital was built with public funding. (MacDonald, 2000)

Similarly, 'another issue is the underestimation of demand risk, which arises if serviced schools or prison places under a 30-year PFI project turn out to be in excess of requirements' (Heald, 2002). Further, Shaoul argues that 'PFI, by locking management into a particular form of service delivery and one contractor for 30 years, serves to reduce rather than enhance management's flexibility to respond to changed circumstances, as other analysis has shown' (Shaoul, 2005, p.10). Thus it can be readily argued that the PFI arrangements reduce flexibility, and inhibit responses to changing circumstance. It can then be said that in that regard PFI arrangements increase overall risk rather than diminish it.

In this discussion, the distinction between risk and uncertainty should also be borne in mind. Risk is taken as the situation where the probability distribution of future 'events' is known (or believed to be known) from which expected outcomes and the variance etc. can be calculated. Uncertainty refers to a situation where the future is unknowable, and where past frequency distributions are not adequate guides to future probability distributions. Further, it may be argued that risk can be applied to 'natural' disasters, e.g. the risk of flooding, but even there, as the example of flooding in the face of climatic change vividly illustrates, past frequencies may not be a guide to future probabilities. The situation becomes more complex when human behaviour is involved: the probability of a contractor fulfilling the contract is not just a function of 'natural events' but of human behaviour and intentions.

Shaoul goes further and argues:

> But risk transfer is conceptually flawed. The concept of risk assumes that all possible outcomes of each trial or event can be predicted and weighted so that a complete array of results covering all eventualities can be compiled. In the context of business decisions, since the number of possible outcomes is infinite, the issue is uncertainty not risk. The significance of this distinction is that it renders the measurement and methodology of risk transfer problematic. (Shaoul, 2005, p. 13)

7. Are costs lower under PFI?

It has been indicated above that in the process leading to the adoption of a PFI project, a comparison has to be made between the proposed PFI and a 'public sector comparator' (PSC). The claims that the PFI projects are more cost effective than 'conventional' public sector projects have been examined by reference to the comparisons between PFI and PSC. In making use of these comparisons the obvious problem arises: the PFI schemes which proceed are those believed to be more cost effective, whereas there are some potential PFI schemes that do not proceed as they are judged to be less cost effective. But, further, the PSC are hypothetical alternatives, and there is generally an incentive for those involved to skew the comparisons in favour of PFI simply because PFI is often seen 'as the only game in town': securing funding and government approval is likely to be easier for a PFI than for a 'conventional' public investment.

The UK government has recently recognized that relatively small projects may not be as suitable for PFI as large projects.

> Similarly, in addition to their own internal costs, bidders must typically meet the costs of technical, financial, design and legal advisors. These costs do not necessarily fall in proportion to the size of the project, and so drive up the relative cost of small PFI schemes. For example, one private sector contractor has suggested that their bid costs, as a proportion of a project's capital value, are 33 per cent lower for a £50 million project compared to a project costing £20 million. (HM Treasury 2003a, p. 53)

It argues that PFI is unlikely to be the best procurement option for projects with small capital value (for example, HM Treasury, 2003a, p. 87).

However there are pressures from central government for the use of PFI, and PFI is often seen as the only route through which a large investment project can be undertaken. There are then clear incentives to overstate the costs of the PSC alternatives in order for the PFI to be favoured. Further, the PFI is carried through whereas the PSC is not, making a genuine comparison fraught with difficulties. It is recognized that

> [t]here is a demonstrated, systematic, tendency for project appraisers to be overly optimistic. This is a worldwide phenomenon that affects both the private and public sectors (Flyvbjerg, *Underestimating Costs in Public Works Projects – Error or Lie*, APA Journal, 2002). Many project parameters are affected by optimism – appraisers tend to overstate benefits and understate timings and costs, both capital and operational.
>
> To redress this tendency, appraisers should make explicit adjustments for this bias. These will take the form of increasing estimates of the costs and decreasing, and delaying the receipt of, estimated benefits. Sensitivity analysis should be used to test assumptions about operating costs and expected benefits. (HM Treasury, 2003b, paras. 5.61, 5.62)

The comparisons between a proposed PFI and a PSC alternative are much influenced by three factors, namely the rate of discount used, the way in which risk is priced and the assumptions made on the project costs of construction. The impact of the way risk is priced has been discussed above.

The judgement between PFI and PSC is influenced by the rate of discount chosen. The PFIs clearly involve financial commitments over periods of up to 30 years as would the PSC. But the time profiles of the costs involved are rather different – the PSC is 'front loaded' with the capital costs incurred in the initial stages, whereas for the PFI as far as the government is concerned the costs are more evenly spread out as the leasing charges repay the capital costs of the constructor. In view of these marked different time profiles, the comparison between PSC and PFI may be rather sensitive to the choice of discount rate. As some of the examples cited in this chapter suggest, the difference in NPV terms between a PFI and the corresponding PSC may be rather small, and hence a relatively small change in the discount rate could well lead to a change in the relative ranking. Further, it may here be noted that the government has recently lowered the test discount rate from 6 per cent to 3.5 per cent, and many PFIs had used the higher 6 per cent rate in the calculations.

Advocates of PFI argue that the project costs will be lower under PFI than under 'conventional' public investment. They would point to a history of cost over-runs on public funded projects, and the incentives which the PFI providers have to ensure that projects are delivered on time and on budget. The strongest arguments for the cost effectiveness of PFI have come from studies such as Arthur Andersen and Enterprise LSE. They report, for example, that

> The total estimated saving from our sample of projects is over £1 billion in NPC terms against an estimated cost of conventional procurement of £6.1 billion. The table shows a consistent pattern of PFI projects delivering sizeable estimated cost savings. The average percentage saving for this sample of 29 projects (i.e. the percentages above added and divided by the number of projects, a calculation that avoids the large projects distorting the average) is 17%. This compares to the average saving in PFI projects with a PSC examined to date by the NAO [National Audit Office] of 20%. On the basis of the public sector's own figures, the data therefore suggests that the PFI offers excellent value for money. (Arthur Andersen and Enterprise LSE, 2000 Section 5.5)

These types of conclusions have been used to justify the inflation of the PSC figures when they are used in the comparison with the proposed PFI. The National Audit Office found, for example, in their report on a GCHQ project that

> [t]he estimated basic construction costs in the final Comparator were increased by 24 per cent in line with Treasury advice on historical cost overruns on large scale public sector projects.... As in other PFI cases, the adjustment for risk on construction costs of the public sector alternative more than accounts for the estimated cost difference between the comparator and the PFI deal. (National Audit Office, 2003, p. 25).

Those who have scrutinized these claims for lowering construction costs under PFI have been much more sceptical of the claims. Pollock, for example, notes the claims which are made

> that PFI projects come in on time and on budget. The Treasury claims that this is true of nearly all PFI projects, whereas most public projects are late and cost more than expected. But researchers at the University of Edinburgh's Centre for International Public Health found that the evidence the Treasury produces for these assertions is either non-existent or false.
>
> The claim that public-sector schemes have average cost overruns of 73%, and time overruns of 70%, is constantly repeated to support the claim that PFI is value for money. But on closer examination it transpires that the only figures the government is willing to release derive from false data commissioned by the government from the PFI industry.
>
> But of the five studies cited by the Treasury as proof of PFI efficiency, only one contains any data. Two reports by the National Audit Office were based on interviews with managers of PFI projects, and the authors themselves conclude that it is not possible to judge from such evidence how the method of procurement affected the results. A third study by a private company contains no comparative data to support the claim. A fourth, by the Treasury, remains under wraps and repeated freedom of information requests have been refused on the grounds that 'disclosure would be detrimental to the commercial interests of the specific PFI contractors'. (Pollock, 2007)

The report by Mott MacDonald which does contain data uses it in a way which does not compare like with like.

The basis of this argument is that there have been cost overruns on past 'conventional' public investment projects but that similar overruns either do not occur under PFI or if they do the consequences are absorbed by the contractors rather than the government. It can be first noted that for some of the examples given above cost overruns do occur under PFI arrangements. As the method of calculation of the 24 per cent figure cited above has not been revealed, we cannot be sure that a like-for-like comparison is being undertaken: the costs of public investment can be affected by inflation, changes of specification, etc. But the point here is that the main argument for lower costs under PFI arises from the nature of the contract. The PFI contract is in effect a fixed price contract which can also contain bonuses for early completion and cost penalties for late completion. These features could readily be incorporated into 'conventional' public investment contracts, which would reduce (or remove) any cost advantages of the PFI arrangements. But any form of fixed price contract and/or contract which involves penalties for late delivery would face difficulties in the case of major unique and innovative projects where forecasting costs ahead of time are fraught with difficulties.

8. Contractual issues

A public private partnership must involve a contract between the public sector and the private sector firm involved. There are a range of well-known issues which arise in the writing and monitoring of contracts and contractual relations. A full analysis would require the comparison between different contractual relationships. For example, in the context of PPPs, an alternative to such a partnership would be for the public sector to carry out the operation completely and this would also involve a series of contracts, which would include employment contracts as well as contracts with suppliers. Contracts involve substantial costs in their writing, specification and negotiation and the monitoring of the contract when in place. Further the way contracts are written provide incentives to behave in particular ways.

The provision of any service involves contractual relations between the parties involved. If a public service is to be provided then some form of contractual relationship must exist between the government and those providing the service. The contracts may be in the form of employment contracts between government and its employees when the public service is provided 'in house', or it may be in the form of a contract between the government and a private sector body (and in turn that body will have employment contracts with its employees and other contracts with its suppliers).

There is now a vast economics literature on contracts, and we pick out some of the salient features of that literature in respect of the PFI and PPP.

It is recognized that contracts involve costs (to all the parties involved) which have been included under the general heading of 'transactions costs'. These costs include the costs of negotiating the contract, the costs of monitoring the implementation of the contract.

In a world of complexity, uncertainty and bounded rationality, contracts are inevitably incomplete. It is simply not possible to envisage and list all possibilities or eventualities which may occur and how the parties to the contract would respond. Employment contracts, for example, may provide general indications of what types of tasks an employee is to undertake, to whom the employee reports, etc. But it is not feasible that an employment contract could specify what an employee is to do each minute of the day, and how the employee would respond in all circumstances.

Contracts often involve a 'principal–agent' situation – that is, where one person (the agent) undertake activities to 'satisfy' another person (the principal). The principal wants certain objectives to be pursued, but for these objectives to be worked towards by the agent. However, the principal will find some difficulty in precisely defining her objectives and ensuring that the agent shares those objectives. The contract which is agreed by the principal and the agent will set up incentives for the agent to behave in a certain type of way. For example, a contract that specifies payment by amount produced creates incentives to produce as much as possible without regard to the quality of what is produced.

The introduction of PFI, PPP and similar forms of contracting out activities which in earlier times would have been undertaken within the public sector may have the effect of pushing the government towards a clearer statement of its objectives. If the contract between the government and the private sector body includes targets to be met by the provider, then it ensures that some consideration is given to the targets to be specified.

The literature on contracts would suggest that fully specified contracts are never feasible. In particular, a contract cannot specify what is to be done in every possible eventuality. On the one side, this limits the ability of the government to specify what is to be undertaken through the contract, especially when the contract extends for many years into the future. On the other side, it generates difficulties for the contract to be flexible and to adjust to changing circumstances.

Contracts involve transaction costs and monitoring costs. In general, a contract is incompletely specified. Indeed if the contract specified in exact detail what the employee was to do in each conceivable circumstance it would rule out any decision making by the employee and any exercise of discretion, knowledge and skills. Further, such a contract would be extremely lengthy, perhaps infinite, in seeking to specify all eventualities and how to respond to them. Apart from the costs involved in writing such a contract, issues of bounded rationality and uncertainty come in: bounded rationality would preclude an individual being able to process all the information, and uncertainty on the future would mean that it is not possible to envisage all possible eventualities. It is also the case that there is generally a situation of information asymmetry in that one party to a transaction has more or better information than the other party: typically the seller knows more about the product than the buyer. The employer and the potential employee have different sets of information: the employer may have more information on what exactly the tasks of a job entail, the employee has more information on her own abilities, apptitudes, attitude to work, etc.

This brief review raises issues for PFI projects. First, there are likely to be substantial costs involved with the contracts, which after all have to extend over 30 years. The Public Accounts Committee note the lengthy time taken in the tendering process. 'The average tendering time for projects in 2004–2006 was 34 months, compared to 33 months for projects that closed prior to 2004. The average cost of advice was £3 million, reflecting the length of the process, and delays to projects cost the taxpayer at least £67 million' (House of Commons Committee of Public Accounts, 2007a, p. 5). A further example is given by Barnes (2001):

> Another key feature of PFI schemes is that it takes a long time to negotiate deals to a successful conclusion. The evidence from recently completed NHS hospital schemes is that it takes approximately 30 months to move from advertisement to contract closure. The process is shorter in local government but the average time taken to conclude deals from the point of advertisement is still 26 months. As this is preceded by a period of investment appraisal and followed by a 12-month to 2-year construction period, it is possible that purchases will have to wait three to four years (at a minimum) before schemes are operational. They should therefore guard against unrealistic expectations and recognize that the PFI is unlikely to be a 'quick fix' solution to their investment needs.

It is, of course, the case that 'conventional' public investment would also involve tendering and negotiations, but in the nature of the long-term contract involved with PFI the costs involved, particularly in the specification of the contract, are likely to be significantly higher.

Second, the length of contract under PFI raises issues of the flexibility of those contracts, as pointed out above. Further, once the contract has been signed, the contractor is in a strong position when there is any requirement to vary the contract in terms of whether to agree to changes and the prices to be charged. This is illustrated in a report from the Public Accounts Committee, when it concluded that

> Departments need to pay very close attention to maintaining value for money throughout the PFI contract period. The NAO census identified that one in five public authorities had already asked for additions or changes to facilities within a few years of letting a PFI contract. Yet in less than half of these cases had the authority attempted to benchmark the resulting price change. A PFI consortium is in a strong position once a contract has been let because it is the contractual supplier for 30 years or more. It is essential that departments take steps to rigorously test any prices for additional work and to impose credible conditions that will allow them to have additional work carried out by alternative suppliers of their choice if there are doubts about value for money. (House of Commons Committee of Public Accounts, 2003b, p. 4)

The National Audit Office reached a similar conclusion in a recent report. It noted that

> [a]lthough public sector requirements are specified in contracts, it is inevitable over the course of 25 to 30 years of operation that changes will be needed to the services and assets provided. It is therefore important that PFI contracts are able to provide the flexibility required at a cost that represents value for money. (National Audit Office, 2008, p. 4)

It continues by pointing to problems and costs with changes to the services agreed in the original contract. These include that 'higher value changes were not always competitively tendered, partly because of timing, cost and the difficulties of integrating new work with the existing set of obligations under a long-term contract, though other provisions to validate costs were sometimes in place.... [Further] for minor works, there was little consistency in the methods used by public sector teams to validate costs and in some instances there was no validation' with wide variations between projects in the charges made for the same minor piece of work (National Audit Office, 2008, p. 4).

These are examples of the inflexibilities introduced within PFI schemes, though the costs of responding to major changes would be much greater than these relatively minor variations. For example, the incorporation of major technological changes in health care or the impact on school provision of demographic changes would involve very substantial costs.

Third, the rationale for schemes like the PFI often involves invoking the benefits which come from competition among potential suppliers. Competition and contract costs have a two-way relationship in the sense that when there is competition between potential contractors there would be increased costs of tendering and contract negotiations involving the successful and the eventually unsuccessful firms. But those contracting costs may well discourage firms engaging in a bidding process. The Public Accounts Committee gives some indication of the relatively small numbers of firms who compete for a contract, and the links with tendering costs. They argue that

> [i]f value for money is to be achieved and maintained, it is critical to have a functioning competitive market, yet the proportion of projects attracting only two viable bids has increased in recent years. Whereas 15% of projects that reached financial close in 2003 or earlier had attracted only two bids, this figure had more than doubled to 33% for those projects that closed between 2004 and 2006... Lengthy tendering periods and high bid costs are discouraging bidders and there is a risk that low market interest may become a significant problem for future PFI deals. (House of Commons Committee of Public Accounts, 2007a, p. 7)

9. Conclusions

In financial terms, the Private Finance Initiative in the UK plays a relatively small role compared with the overall volume of public expenditure, but does have a substantial role in the investment programmes in education, health and transport. But PFI should be seen as one important aspect of a programme of privatization and the operation of 'new public management'. It involves the creeping involvement of the private sector in the provision of public services and the ownership of assets used for the provision of public services. In this chapter we have argued that, while concern over declining public investment was a force behind the expansion of the PFI programme, the financing of public investment via PFI rather than through 'conventional' public investment does not release the economic constraints on the scale of public investment. Any higher levels of public investment come from political decisions and decisions could equally have been made to increase public investment through the 'traditional' routes. The manner in which PFI is reported in the public accounts (no addition to present budget deficit, the future obligations not included in public accounts, and adding to future public expenditure) encourages the use of PFI but builds up future obligations. These obligations stretch out over the next 30 years, and involve the public purse in overall greater expenditure than would have been the case with 'conventional' public investment. The cost of finance is clearly significantly higher for PFI compared with 'conventional' public investment. This is widely acknowledged but two arguments have been advanced to counter that higher cost. The first is that under PFI risk is transferred to the private sector and the private sector is better able to deal with that risk. Further, as with an insurance contract, there is a cost associated with the transfer of risk. We have examined this argument and found it wanting. We have cast doubt as to whether risk is effectively transferred and have pointed to the inflexibilities that a PFI scheme introduces. The second argument is that the projects constructed under PFI are more likely to be delivered on time and on budget than under 'conventional' public investment. It has been argued here that the evidence for that assertion is doubtful. But the ways in which risk transfer is priced and attaching higher costs to the PSC (public sector comparator) in the assessment of a proposed PFI project are often crucial to decisions to proceed with the PFI project.

It then has to be concluded that the PFI programme is a rather expensive way of securing higher levels of public investment, which could be readily secured under 'conventional' investment financing.

Note

1. As this chapter was being completed, Northern Rock was being rescued by the UK government, and the terms of that rescue add up to £100 bn to the public debt (though there are, hopefully, corresponding assets being acquired). The debt ratio of 40 per cent is then clearly breached.

References

Arthur Andersen and Enterprise LSE (2000) 'Value for money drivers in the private finance initiative' (from web site www.ogc.gov.uk).

Barnes, M. (2001) 'Putting the PFI into practice', *Public Finance*, 29 June 2001 (on-line version).

Boateng, P. (2001) 'Keynote address' at the Public Private Partnerships/Private Finance Initiative Global Summit, Dublin, 16 October 2001.

Broadbent, J. and Laughlin, R. (2003), 'Public private partnerships: an introduction', *Accounting, Auditing and Accountability Journal*, Vol 16: 332–41.

Broadbent, J. and Laughlin, R. (2005a), 'The Role of PFI in the UK's Modernisation Agenda', *Financial Accountability and Management*, 21(1): 75–97.

Broadbent, J. and Laughlin, R. (2005b), 'Government Concerns and Tensions in Accounting Standard Setting: The Case of Accounting for the Private Finance Initiative in the UK', *Accounting and Business Research*, 35(3): 207–28.

Edwards, P. and Shaoul, J. (2002), 'Partnerships: for better, for worse?', *Accounting, Auditing & Accountability Journal*, 16(3): 397–421.

Falconer, Lord (2001) 'The Politics of Major Projects' Lecture at University of Strathclyde, 22 March 2001.

Grimsey, D. and Lewis, M.K. (2004), *Public Private Partnerships*, Cheltenham: Edward Elgar Publishing.

HM Treasury (2003a), 'PFI: meeting the investment challenge', accessed on www.hm-treasury.gov.uk.

HM Treasury (2003b), *The Green Book Appraisal and Evaluation in Central Government*.

Heald, D. (1997), 'Privately financed capital in public services', *Manchester School*, vol. 65: 568–598.

Heald, D. (2002) *Observer*, 28th April 2002.

House of Commons Committee of Public Accounts (2003a), *Delivering better value for money from the Private Finance Initiative*, Twenty-eighth Report of Session 2002–03, HC 674.

House of Commons Committee of Public Accounts (2003b), *PFI Construction Performance*, Thirty-fifth Report of Session 2002–03 HC 567.

House of Commons Committee of Public Accounts (2003c), *Private Finance Initiative: Redevelopment of MOD Main Building*, Fourth Report of Session 2002–03, HC 298.

House of Commons Committee of Public Accounts (2004), *Government Communications Headquarters (GCHQ): New Accommodation Programme*, Twenty-third Report of Session 2003–04, HC 65

House of Commons Committee of Public Accounts (2007a), *HM Treasury: Tendering and benchmarking in PFI*, Sixty-third Report of Session 2006–07, HC 754.

House of Commons Committee of Public Accounts (2007b), *Update on PFI debt refinancing and the PFI equity market*, Twenty-fifth Report of Session 2006–07, HC 158.

MacDonald, R. (2000) 'Private finance initiative condemned', *British Medical Journal*, 16 September, 2000, Vol. 321: 657.

National Audit Office (2002), *PFI refinancing update*, HC 1288 Session 2001–02.

National Audit Office (2003), *Government Communications Headquarters (GCHQ): New Accommodation Programme*, HC955 July 2003.

National Audit Office (2008), Making Changes in Operational PFI Projects HC 205 Session 2007–08, January 2008.

NHS Confederation (2004), *Getting the best out of future capital investment in health*, London: NHS Confederation.

Pollock, A.M., Shaoul, J. and Vickers, N. (2002) 'Private finance and "value for money" in NHS hospitals: a policy in search of a rationale?', *British Medical Journal*, Vol. 324: 1205–9.

Pollock, Allyson (2007), 'A gauntlet for Brown: Treasury claims that PFI offers value for money are based on data that is non-existent or false', *The Guardian*, 11 April, 2007.

Shaoul, J. (2005), 'A critical financial analysis of the Private Finance Initiative: selecting a financing method or allocating economic wealth?', *Critical Perspectives on Accounting*, 16(4): 441–71.

Sussex, Jon (2001) *The economics of the private finance initiative in the NHS*, Office of Health Economics: Summary on www.ohe.org (accessed 9 May 2002).

Timmins, Nicholas (2007), 'As spending tightens capital may run short', *Financial Times*, 27 July, 2007.

Unison (2004), 'Public risk for private gain? The public audit implications of risk transfer and private finance', report research and written for UNISON by Allyson Pollock, David Price and Stewart Player.

3
Water Privatization

David Hall and Emanuele Lobina

University of Greenwich, UK

Abstract

The chapter examines the expansion of private water companies since 1989, the withdrawal from developing countries from 2003 onwards and the economic impact of privatization. The analysis is set in the context of the historical development of water services in the north and the south, showing that the role of private water companies since the start of the twentieth century has been historically limited and exceptional. The impact of water privatization is considered in relation to the issues of investment, prices and efficiency, drawing on empirical evidence from the north and developing countries in Asia, Africa and Latin America. Particular attention is given to France and the UK where private water companies, for different reasons, are most established. The evidence from both north and south shows systematic underinvestment, monopoly pricing, regulatory gaming, and no significant efficiency differences between public and private sector operators. In conclusion, the chapter identifies institutional policies including fiscal constraints and lending conditionalities as key drivers of privatization, and questions whether these can sustain privatization in the water sector where historical experience indicates it is an inappropriate solution.

Keywords: privatization, water industry

Journal of Economic Literature classification: L1, L33, L95

1. Introduction[1]

The introduction of private companies into water and sanitation services over the last 20 years has provoked considerable political, social and academic debate. It has involved major multinational companies and international institutions, consumers, trade unions and social movements. This experience and debate has taken place in nearly all countries, both north and south. This chapter locates the recent trends in the historical context of the development of public water services in the nineteenth and twentieth centuries. It then examines the evidence on economic features of the recent privatizations, in terms of investment, prices and efficiency. It then discusses the conclusions that can be drawn from this experience, and argues that these mirror the conclusions drawn in the development of public water services a century ago.

This chapter is constructed in five main sections, followed by a conclusion.

- It discusses the historical development of the role of the private and public sector in water.
- It provides an account of the development of private water companies from the 1980s, and the institutional and political context in which this took place.
- It assesses what private water companies have contributed to investment in extending access to water and sanitation systems.
- It considers evidence on whether private water companies extract monopoly profits.
- It reviews evidence on whether private water companies are more efficient than public water companies.
- The conclusion then draws parallels between the experience of the last 20 years and the factors that led to the growth of the public sector a century ago.

2. History: the dominance of the public sector

The history of the development of water and sanitation systems in the high income countries of the north shows a common pattern. In Europe, urban water systems began developing in the seventeenth or eighteenth centuries as a limited service to affluent customers and as a public assistance for fire control. As cities grew in the nineteenth century, the demand for water consumption grew and the public health issues became more acute. While the initial systems were usually started by private companies, during the nineteenth century the utilities were fairly soon taken over by municipalities in nearly all European countries, including the UK. Only in France did the old nineteenth-century private operators survive, which is why the only large water companies in the world are French: Suez (formerly Lyonnaise des Eaux) and Veolia (earlier Vivendi and the Compagnie Générale des Eaux) since 1853.

Municipalization was seen as a way to overcome the systemic inefficiencies of the private contractors:

> During the 19th century, the previously private systems came under public ownership and public provision because of the inefficiency, costs and corruption connected to them.... Democratically elected city councils bought existing utilities and transport systems and set up new ones of their own. This resulted in more effective control, higher employment, and greater benefits to the local people. Councils also gained the right to borrow money to invest in the development of their own systems. (Juuti and Katko, 2005)

This was linked to the growth of municipal socialism (or 'gas and water socialism'), which saw the public sector as a mechanism to fulfil a set of economic and political objectives – economic development, public health and improvement of social conditions for the urban poor. Public finance mechanisms were similarly central to the development of water systems in the USA. Up to the 1880s most American cities had water and sanitation systems that were inadequate in terms of public health, fire risk and social and economic development. By the 1930s the majority of cities had developed comprehensive and reliable systems, largely under municipal control. The municipalities developed financial mechanisms, superior to the private sector, including borrowing long-term money from local savers, at low interest rates because of the security of their flow of income from taxes (Melosi, 2000; Cutler and Miller, 2005).

Despite the dominant role of municipalities, central governments have also played a significant role in financing water systems. This has sometimes involved paying directly for the water supply service, so that there is virtually no role for charges (e.g. Ireland); distributing some part of central tax revenue to support local authority spending on water and other services (e.g. Canada); providing cheap loan finance for local authorities to use for capital investment (e.g. USA); or collecting part of the water charges centrally and redistributing it to authorities that need to invest (e.g. France). In Europe, the EU itself plays a major role in public financing of water systems in poorer states, and through low interest loans from its public sector development instrument, the European Investment Bank. France and the UK are the only two OECD countries whose water operations are now mostly run by private companies. However, in both countries the cost of extending water and sanitation networks

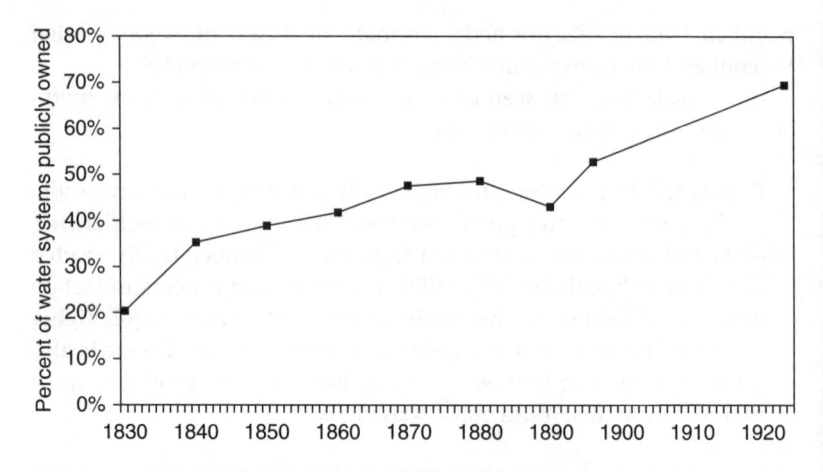

Figure 3.1 Public ownership of water systems in USA cities 1830–1924
Sources: Cutler and Miller, 2005; Galishoff, 1980; Melosi, 2000.
Note: Data available on the decade from 1830 to 1890 and for 1896 and 1924; all points in between are obtained by linear interpolation.

has been met through public finance mechanisms. The case of France is of special interest as it is the home country of the major private water companies, which have operated since the mid-nineteenth century. Despite this, their contribution to investment for extension of the system has been negligible, and the development of the system in France has depended on public authorities using taxation and cross-subsidies. In the nineteenth century private companies were given concessions for providing a water supply to public taps and fountains. However, there was no universal service obligation on the companies to provide piped water concessions to every household, and so the companies could be selective about whom they chose to serve. By 1900 only 2 per cent of French households had direct connections, and the municipalities were unable to accelerate development through the private companies. In order to provide a universal service, municipalities had to finance investment themselves. By the end of the 1930s, 32 million people were supplied with tap water, supported by public finance rather than operating surplus: 'Urban local authorities financed the development of the public service themselves' (Pezon, 2007). In the 1950s a similar solution was used to finance connection in rural areas. The National Fund for Rural Water Supply (FNDAE) was created in 1954 in order to finance the cost of connecting these households, by levying a tax per cubic metre of water on all water supplied in France. The money was then

Table 3.1 Types of water system and connections to the system in France

		Dominant type of operator	Investment	Operation	Connection at start of period	Connection at end of period
Period A	1848–1900	Concession	Private	Private	0%	2%
Period B	1900–1970	Régie (municipal)	Public	Public	2%	90%
Period C	1970–present	Affermage	Public	Private	90%	–

Source: Based on Pezon (2003).

distributed to rural communes to finance the necessary investments in constructing new networks and connections. The extensions of the system in rural France were thus financed through a massive cross-subsidy from households and businesses already connected to the system. By the mid-1990s the rural connection rate had reached over 95 per cent (Reynaud, 2006).

Water supply in developing countries has a different history. In the colonial period, while the imperial countries were extending public networks in European cities, water supply in the colonies was focused on a colonial elite. These elite systems left a physical legacy of incomplete networks (Gandy, 2004; Nilsson, 2006).

Colonialism also left a socio-economic legacy of more unequal societies, which makes both the problem more acute and the requirement for redistributive public finance greater. After independence, it was possible to start developing the physical and social infrastructure of public services for all. The commitment to water supply and other public services was thus closely associated with the process of building independent states with political accountability to their citizens for the first time.

In many developing countries central government has played a greater role in the water systems than in the north. Driven by independence rather than industrialization, these countries had neither strong municipalities nor a strong local middle class, and so central state ownership of water providers is more common than in the north. In Sri Lanka, a country with an excellent developmental record on health and education, water has been primarily the responsibility of a central government parastatal. In a number of countries, including Uganda and Honduras, central government has retained ownership of the capital city's water operation, which has then been used as an agency to support development of municipal services elsewhere. In Argentina, the extensions of water systems throughout the country were carried out by a central government water agency. But development in non-industrialized countries has continued to be strongly affected by the economic and political demands of international agencies and donors, and water services are a clear example of this. The IMF and World Bank conditions of the early 1990s nevertheless insisted on making municipalities responsible for services, facilitating the break up and privatization of Argentina's previously national system: there have been continuing pressures on Sri Lanka to do likewise (Castro, 2004; Mycoo, 2005).

Despite the expansion of the 1990s, water supply services remain overwhelmingly dominated by the public sector. Around 90 per cent of the 400 largest cities in the world, with populations of over 1 million, are served through public sector operators. The proportion is around 85 per cent for these largest cities in high income countries, including western Europe and north America: some estimates for the water sector in Europe offer a figure as low as 55 per cent, but only by treating as 'private' cooperatives and operators controlled by the public sector but with some minority private shareholding (Hall and Lobina, 2007a; Euromarket, 2004). In Africa and Latin America the proportion is also around 85 per cent, while in south and east Asia it is over 90 per cent.

Private or public water operators in cities with population over 1 million (October 2006)

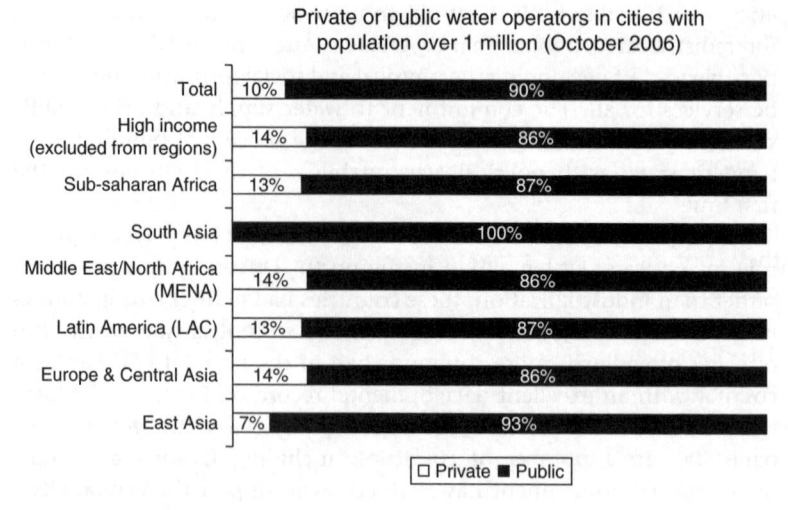

Region	Private	Public
Total	10%	90%
High income (excluded from regions)	14%	86%
Sub-saharan Africa	13%	87%
South Asia	0%	100%
Middle East/North Africa (MENA)	14%	86%
Latin America (LAC)	13%	87%
Europe & Central Asia	14%	86%
East Asia	7%	93%

Figure 3.2 Public and private ownership of water operators, major cities, 2006
Source: Hall and Lobina, 2007a.

3. Water companies

The history makes clear that by the 1980s the great majority of water supply and sanitation networks in the world were in the public sector. Private sector activity in water and sanitation services fell into three different categories, all of which can be characterized as residual.

First, the French water companies, which had survived waves of municipal expansion and nationalization. By the end of the 1980s they had grown to dominate the provision of water services in France, with major public works construction divisions, and developing increasingly strong positions in other public services including waste management, heating and energy services, and health care (Barraque, 1995; Pezon, 2006). No other country had any companies comparable to this French group – the remaining Spanish and Italian private water companies were partly owned by the French companies – until the privatization of the English and Welsh companies in 1989. This was a political decision by the Thatcher government, made possible because England and Wales, uniquely in western Europe, had restructured its water sector 15 years earlier, so that all municipal operations had been merged into a small number of state-owned regional companies. All the expansion in privatization in the 1990s involved this small group of French companies, and, to a much smaller extent, some of the English companies.

Secondly, some independent private companies also survived municipalizations, but represented only 10–15 per cent of the sector, in countries such as Germany, the UK, and the USA, typically regulated with a low but secure rate of return. None of these expanded in the 1990s, although a number of them were taken over by expanding multinationals. Thirdly, in developing countries, large numbers of small street vendors and kiosk operators supplied water to those without connections or access to a public piped water service. These vendors thus operate in markets which are defined by the failure of public water services, and they continue to play a large part in selling water to those without a reliable public supply.

3.1 Expansion

Two political factors were important drivers of the subsequent expansion of water privatization. The first was the ideological change of the 1980s, symbolized by the privatization of water in the UK, demonstrating that it was both possible and potentially profitable to privatize water. The second was the strategy adopted by the World Bank and donor agencies to promote water systems in developing countries through privatization.

This was expected to deliver finance for investments, efficiency improvements, and better governance than they believed possible through the public sector in developing countries. It was expected that multinational companies would be attracted by a large new profitable market, and that the process would be welcomed by populations disillusioned with the corruption and inefficiency that the World Bank associated with the public sector. It was so central to donor policies that a World Bank official told an international conference in 2000 that 'there is no alternative' to privatization.

The expansion of the private companies in the 1990s was global in scope, and from 1990 up to 2003 the global share of private water operations grew at a considerable pace, though still remaining a small minority. The attempts at expansion into north America and other west European countries had little success except in the two countries where the French multinationals were already established – Spain and Italy – and to a limited extent in the USA. In Germany for example the private companies were only able to win a few concessions in former east German towns such as Rostock and Potsdam, apart from Berlin itself. Expansion in developing countries was initially far more successful, but has also been reversed since 2002 in the wake of political opposition and failure to make economic returns. The most sustained expansion took place in the former communist countries of central and eastern Europe, where over 35 major cities or regions remained in the hands of the multinationals at the end of 2007 (Hall and Lobina, 2007b).

The French companies dominated this expansion. At the peak of water privatization in around 2002, Suez (whose water division has also been known as Ondeo, and originally as Lyonnaise des Eaux) and Veolia (previously part of Vivendi, and originally known as Generale des Eaux) shared 60 per cent of the 320 million customers served by multinationals (Hall and Lobina, 2007b). SAUR was also involved, especially in Africa and Europe. Part of its expansion involved takeovers of English companies, mainly the smaller ones. In Latin America, Suez used its Spanish affiliate Aguas de Barcelona as a lead partner; Veolia used a similar approach, first buying a half stake in the Spanish group FCC, and then setting up a joint venture, Proactiva, to pursue water opportunities in Latin America.

Most of the English and Welsh companies attempted to expand internationally, including Hyder, Severn Trent, Anglian, Yorkshire, Thames, and United Utilities – the latter initially in partnership with Bechtel, the large USA construction company. By 2006 all of these had retreated, except for United Utilities. A privately owned British construction company, Biwater, bought a small English water company and sought international business. Two energy multinationals attempted to enter the market, both of which did so by buying English water companies. The USA group Enron bought Wessex Water, formed a water division Azurix, which failed; the German energy group RWE bought Thames Water, which became the third largest water company in the world, before selling its international operations and then being sold by RWE to an Australian finance group, Macquarie.

One consequence of the dominance of the public sector was that the private companies were not growing by competing among themselves. The growth had to come by making inroads into the services provided by the public sector, and the great majority of public sector operators did not seek to compete with the private companies by expanding. Moreover, the total number of private companies seeking to grow was very small, thus forming a de facto oligopoly, which often formed joint ventures with each other. This was reinforced by the fact that the existing companies were protected against new entrants by the length of their existing concessions, lasting 25 or 30 years and in some cases much longer.

Since the target market was in the public sector, political decisions were necessary to enable private companies to expand in the sector. It was thus unsurprising that the companies' growth in all continents was characterized by close relationships with development banks – especially the World Bank – donors and politicians. For the French companies, this

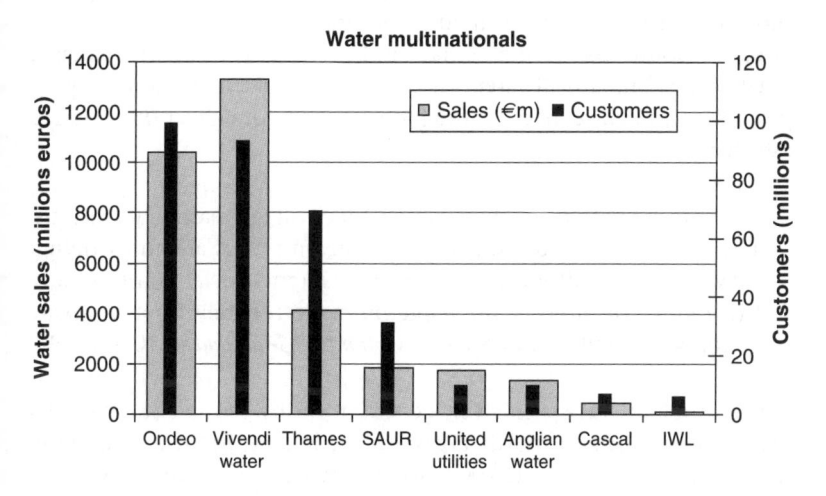

Figure 3.3 Multinational water companies in 2002
Source: Hall and Lobina, 2007b.

was an extension of their intense, and sometimes corrupt, relationships with politicians which had facilitated their own survival and growth in France (Hall and Lobina, 2007b).

3.2 Problems

Since 2002 all the multinational groups with water divisions have been seeking to leave or reduce their stake in the water sector. This trend is visible world-wide, as companies withdraw from developing countries, and as private companies are sold to new owners, often financial groups. Three reasons can be identified for this withdrawal.

First, and fundamentally, the multinationals failed in general to make an acceptable return for their shareholders. Suez issued a statement in January 2003 stating that it was withdrawing from developing countries, and would not in future invest in any operation which was not both self-financing and delivered an acceptable return, free of currency risk: 'Suez' exposure to emerging countries, as measured by capital employed, is expected to be reduced by close to one third'. The problems included currency devaluations, economic crises, over-optimistic projections, and public resistance to price rises. But they also included the same problems that the companies had encountered a century before in Europe and north America, namely the impossibility of making profitable investment in extensions and improvements for poor households who were unable to pay the full cost of water supplied, without substantial public subsidy. A selective service could be profitable, but not a universal service. The point was succinctly made in December 1999 by the manager of UK water company Biwater, which pulled out of a major water supply project in Zimbabwe, because the project could not deliver the required rate of return:

> Investors need to be convinced that they will get reasonable returns. The issues we consider include who the end users are and whether they are able to afford the water tariffs. From a social point of view, these kinds of projects are viable but unfortunately from a private sector point of view they are not. (*Zimbabwe Independent* 10/12/1999)

The problem was specified in general form by J.F. Talbot, the chief executive of SAUR, the fourth largest water company in the world, speaking to the World Bank in 2002. He referred to the huge scale of the needs, acknowledged that the extension of water supply was necessary for sustainable development, but openly asked 'is it a good and attractive business?'. He rejected the assumption that the private sector could raise funds on the necessary scale, criticizing:

> [a]n often premature or simply unrealistic emphasis on concession contracts and full divestiture... A belief that any business must be good business and that the private sector has unlimited funds.... The scale of the need far out-reaches the financial and risk taking capacities of the private sector.

He warned that tighter contracts and regulation make things worse from a business perspective: the general increase in risk was made worse by: 'Unreasonable contractual constraints.... Unreasonable Regulator power and involvement'. And there was also '[a]n emphasis on unrealistic service levels... Attempts to apply European standards in developing countries.... The demand for "connections for all" in developing countries'. Finally, he rejected the possibility of cost recovery from users: 'water pay[ing] for water is no longer realistic in developing countries: Even Europe and the US subsidise services.... Service users can't pay for the level of investments required, not for social projects...'. The solutions to these problems, in his view, was for public sector subsidies, soft loans and guarantees, without which the multinationals would withdraw:

> substantial grants and soft loans are unavoidable to meet required investment levels... The considerable dependence of the growth of the water sector in the developing world on soft funding and subsidies. If it does not happen the international water companies will end up being forced to stay at home. (Talbot, 2002)

Secondly, there has been a remarkable degree of public and political opposition to water privatization. This has been visible in campaigns globally, in both north and south. The opposition includes trade unions, environmentalists, consumer groups, citizens' organizations, elected politicians and other groups. A common theme of opposition campaigns include the belief that water supply is an essential service, which should be public, that companies should not be allowed into a position where they can profit from their monopoly of vital resource; another is a reaction against what is usually perceived as a foreign private capture of a vital national service, and resentment of the imposition of conditionalities by the World Bank and IMF. The uprising which led to the termination of the private water contract in Cochabamba (Bolivia) in 2000, was the first and most dramatic of a series of reversals: in 2004 another uprising in

Cost of equity vs. RoE by sector

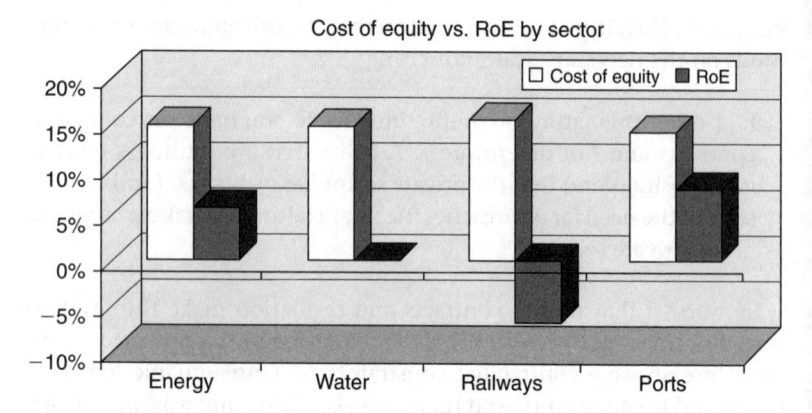

Figure 3.4 Returns on infrastructure investment in developing countries
Source: Estache and Pinglo, 2004.

El Alto, the poor suburb of La Paz, led to the termination of Suez's concession in that city. Over 71 per cent of people strongly supported the renationalization of the water service of Buenos Aires in 2006, according to an opinion poll. The unpopularity of privatization is such that two countries in the world – Uruguay and Netherlands – have made water privatization illegal (Hall and De La Motte, 2004; Hall *et al.*, 2005).

The opposition was reinforced by evidence and suspicion of corruption not only on the part of politicians and officials receiving bribes, but also by western multinationals offering bribes. This involved not only using inducements to obtain specific contracts, but also attempts to obtain control of the policy-making process itself through a process of 'state capture', through donations that may be legal in some contexts (Hall 1999; Hellman *et al.*, 2003; Kaufmann and Vicente, 2005).

Even in the UK, where it is often assumed privatization has widespread public support, after 17 years of water privatization, a clear majority of 56 per cent favour a return to public ownership, according to the results of an opinion poll in June 2006. This represents a continuation of consistent public opposition to water privatization, apparent throughout the 1980s when water privatization was being proposed and introduced. The first proposals in 1985 were widely criticized: even a *Financial Times* editorial suggested that: 'the water industry has many special characteristics which seem to justify public ownership'. A poll in December 1986 showed that 71 per cent were opposed to water privatization, by December 1988 the majority against water privatization had risen to

Table 3.2 Opposition to privatization of water: some world-wide examples, 1994–2002

Year	Country	City	Event
1994	Poland	Lodz	Privatization proposals rejected
1995	Hungary	Debrecen	Privatization proposals rejected
1995	Sweden	Malmo	Privatization proposals rejected
1996	Argentina	Tucuman	Termination of privatization
1996	USA	Washington DC	Privatization proposals rejected
1998	Germany	Munich	Privatization proposals rejected
1999	Brazil	Rio	Privatization proposals rejected
1999	Canada	Montreal	Privatization proposals rejected
1999	Panama	National	Privatization proposals rejected
1999	Trinidad	National	Termination of privatization
2000	Bolivia	Cochabamba	Termination of privatization
2000	Germany	Potsdam	Termination of privatization
2000	Mauritius	national	Privatization proposals rejected
2000	USA	Birmingham	Termination of privatization
2001	Argentina	BA Province	Termination of privatization
2001	France	Grenoble	Termination of privatization
2002	Brazil	National	Continuing campaign
2002	Ghana	Accra	Continuing campaign
2002	Indonesia	Jakarta	Continuing campaign
2002	Paraguay	All	Privatization proposals rejected
2002	Poland	Poznan	Privatization proposals rejected
2002	S Africa	National	Continuing campaign

Source: Hall et al., 2005.

75 per cent, and to 79 per cent in July 1989. *The Times* commented of these privatization plans: 'by and large the public sees little point and only disadvantages in them. They seem simply doctrinaire.' The water companies were nevertheless privatized a few months later (Hall and Lobina, 2008).

The third major factor was the failure of the private sector to live up to expectations, especially in terms of investment, exacerbated both of the previous problems. Suez's concession in Manila (Philippines) had become the subject of a bitter dispute with the regulator, and by 2006 had been 84 per cent renationalized. The collapse of the Argentinian economy led to the ending of water concessions in Buenos Aires and Santa Fe as the companies failed to force Argentina to guarantee profits in dollars. In Africa, contracts were terminated in Gambia, Mali, Chad, Nkonkobe (South Africa) and Dar-es-Salaam (Tanzania). Privatization has

88 *Critical Essays on the Privatization Experience*

Support for public ownership of water in England and Wales, 2006

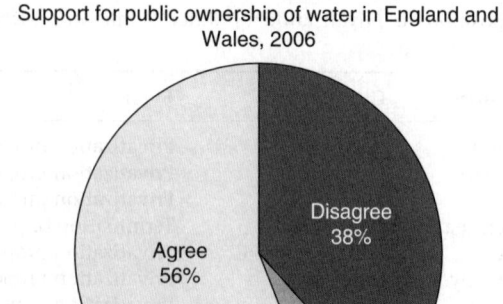

Disagree 38%
Agree 56%
Don't know 6%

Figure 3.5 Popular support for public ownership of water industry in the UK, 2006
Source: BBC Daily Politics Show Poll Fieldwork: 14–15 June, 2006. Conducted by Populus. http://www.populuslimited.com/pdf/2006_06_20_Daily_Politics.pdf

faced similar rejections and reversals in developed countries: in the USA, for example, the city of Atlanta terminated Suez's concession because a public sector operation would be better value.

3.3 Retreat

Both Suez and Veolia have reduced their activities internationally, but broadly maintain their presence in Europe, North America, China, and (for Suez) North Africa and the Middle East. Developing countries are no longer on the map of possibilities.

The specialist water companies have either sold their international operations: Anglian Water, Severn Trent, Thames – or are seeking to withdraw or reduce their exposure as much as possible: SAUR, United Utilities, Berlinwasser, Gelsenwasser. Groups dominated by non-water business have sold their water interests completely, including Bechtel, Bouygues, E.ON, and RWE (largely). The two largest groups have effectively experienced the same process: Veolia Environnement was floated as a water and waste management company by Vivendi, the media multinational which had itself originally grown out of the water and waste business of Generale des Eaux. In 2007, Suez Environnement was also being created as a water and waste company, being separated from its parent Suez, which had originally grown from the water and waste base of Lyonnaise des Eaux, but was now merging with Gas de France and making itself a pure energy company.

The problem has been finding buyers for these water operations. A significant proportion of the new owners are private equity funds, including specific infrastructure funds, and various public sector bodies including governments, municipalities, and state investment agencies. For example, Bechtel's water interests in Europe were up for sale for over a year and in the end were bought by a public development bank, the EBRD. Bouygues' water company, SAUR, the fourth largest in the world, was for sale for two years, before being bought by the private equity firm PAI, who refused to take on the non-European operations, which Bouygues has since sold piecemeal. Thames Water, the third largest water multinational, was formally put up for sale by RWE in November 2005, and finally sold in December 2006 to an infrastructure fund run by Macquarie Bank of Australia. In order to complete the sale Thames was required to sell off its overseas interests.

In April 2007 PAI sold SAUR to a consortium led by the French state bank Caisse des dépôts et consignations (CDC), which holds over 40 per cent of the shares. In effect, SAUR has been partially nationalized. This was done in order to prevent a foreign private equity takeover of the French operations: 'The consortium's offer was chosen not only because it was the best, but also because it will allow the water distribution company to remain French-owned.' (*Les Echos*, 20 April, 2007) This is part of a more general bi-partisan French strategy to create a state capitalism actively ensuring local French control of major private companies in infrastructure, property and health care. In water this strategy is now almost complete. At the start of 2008 the French state, either directly or through CDC, owns more than 33 per cent of the newly autonomous Suez Environnement; and CDC owned 10 per cent of Veolia.

The ownership of the privatized water companies in England and Wales shows a major growth in the role of financial and private equity investors. This has been accompanied by a general withdrawal of equity finance and its replacement by debt financing and private equity. At the start of 2008 only four of the 10 large water and sewerage companies are still quoted on the stock exchange, and of these, Northumbrian is 45 per cent owned by three financial investors, and 30 per cent of Pennon Group, owners of South-West Water, is owned by five major financial shareholders. Five other large companies – Anglian, Southern, Thames and Yorkshire – were already owned by private equity or financial groups by the end of 2007. Only one is now owned by a multinational group – Wessex, owned by Malaysian company YTL; and one is a not for profit company (Glas Cymru). Of the smaller water only companies, three are still owned by Veolia, one is now owned by Suez/Agbar, one by a Hong Kong group, one by a private UK group (Biwater), and the rest are owned by private equity.

The great majority of European water operators remain in public ownership. Among those that have been or remain privately owned, there is no consistent pattern of ownership emerging to replace the multinationals. In some cases public authorities have re-purchased ownership of the water companies (the state in the case of SAUR in France and Elber in Albania, municipalities in the case of Gelsenwasser in Germany); in a few cases local companies have purchased shareholdings from the multinationals (e.g. GW-Borsodvíz in Hungary); and there have also been cases of shares being sold to the public (e.g. Tallinn).

3.4 South

As the multinationals retreat from their international investments, a mixed pattern is also emerging in the south. The difference is that governments and municipalities are the main new owners of formerly privatized water operations. In Latin America, where privatization made greatest advances, there are now a number of water concessions remaining in the hands of private European companies, but all of these owners would prefer to sell the concessions if possible. There is one private equity investor, Ontario Teachers Pension Plan, which now owns a group of Chilean private water companies, but no others. There are a number of cases where local private sector companies have taken over private concessions: it remains to be seen whether this is a significant future pattern, or whether it is just an interim stage in a slower return to public ownership.

In Argentina, the renationalization of water in Buenos Aires re-establishes a strong role for central government in the sector, which was the case before the privatizations of the 1990s were induced. It is noteworthy that workers and unions often have a formal ownership stake in the new public entities. This is the result of the employee shares introduced at the time of privatization, which were originally intended to buy off opposition from workers and unions.

In Brazil, which has a mixture of state and municipal water operators, there is a range of developments. The association of municipal operators, Assemae, has been actively encouraging the development of municipally owned operators, including the use of public–public partnerships. In the other direction, two of the major state-owned companies in Brazil have

Table 3.3 France and UK: water company ownership, December 2007

Company	Principal owner	Country	Type of owner	Comments
Suez Environnement	GdF-Suez	France	SEC	35% owned by GDF-Suez, +5% by CDC.
Veolia Environnement		France	SEC	10% owned by CDC.
SAUR	CDC	France	PE/state	40% owned by state investment agency CDC.
Anglian Water	Osprey/AWG	UK	PE	Consortium of 3 PE funds, inc. 3i.
Northumbrian Water		UK	SEC	25% owned by Ontario Teachers Pension Fund, 15% by fund managers Amvescap, 5% by Barclays Bank.
North West Water	United Utilities	UK	SEC	
Severn Trent Water	Severn Trent	UK	SEC	
Southern Water	Greensands	UK	PE	Main partners are JP Morgan and Challenger. Bought October 2007.
South West Water	Pennon Group	UK	SEC	Pennon is 30% owned by 5 financial investors.
Thames Water	Macquarie	Australia	PE	
Welsh Water	Glas Cymru	UK	NPC	
Wessex Water	YTL	Malaysia	M	
Yorkshire Water	Citigroup/HSBC	UK	PE	Bought November 2007.
Bournemouth and West Hampshire Water	Biwater	UK	P	Private company, operates internationally, but not in EU outside UK.
Bristol Water	Agbar/Suez	ES/FR	M	
Cambridge Water	Cheung Kong Infrastructure	Hong Kong	M	

(Continued)

Table 3.3 (Continued)

Company	Principal owner	Country	Type of owner	Comments
Cholderton Water	Cholderton Estate	UK	P	Private family owned.
Dee Valley	–	UK	SEC	35% of shares owned by Axa SA.
Folkestone and Dover	Veolia	FR	M	
Mid Kent Water	UTA and HDF	Australia	PE	Utilities Trust of Australia (UTA); Hastings Diversified Utilities Fund (HDF). Bought Swan Group, the holding company of Mid Kent Water. Swan also owns 51% of Halcrow water Services.
Portsmouth Water	South Downs Capital	UK	PE	South Downs Capital is 36% owned by SMIF/Land Securities (PE). SMIF = Secondary Market Infrastructure Fund. SMIF itself was bought by Star Fund (PE) in 2003, sold in 2006 to Land Securities (PE).
South East Water	UTA and HDF	Australia	PE	Macquarie bought South East Water from SAUR in 2003; sold it to UTA/HFM in October 2006, prior to o purchase of Thames Water.
South Staffordshire Water	Alinda Capital Partners	USA	PE	Bought in 2007 from Arcapita (Bahrain)
Sutton & East Surrey Water	Aqueduct Capital	DE	PE	Aqueduct Capital is part of Deutsche Bank. Bought holding company East Surrey Holdings Group (ESH) for £189m in 2006 from Kellen Acquisitions Ltd – part of Terra Firma. Kellen had bought ESH only in October 2005, and then sold off gas companies.
Tendring Hundred	Veolia	FR	M	
Three Valleys	Veolia	FR	M	

Notes: Type of owner: State = state; SEC = stock exchange quoted; M = multinational; PE = private equity; NPC = not-for-profit company; P = privately been part-privatized by the sale of shares to investors through the stock exchange. SABESP, owned by Sao Paulo state, is 49.7 per cent owned by investors through the New York and Sao Paulo stock exchanges. Copasa, owned by Minas Gerais state (59.8 per cent) and the municipality of Belo Horizonte (9.7 per cent), is also listed on the Sao Paulo stock exchange, and 30.24 per cent owned by private investors. Both these companies are also engaged in international 'partnerships': SABESP with the utility Sedepal, in Lima, Peru; and Copasa with the Paraguayan state water company Essap.

In Colombia, which has both a multinational and a local private operator, three municipally-owned Colombian water operators are trying to expand into other areas: EAAB (Empresa de Acueducto y Alcantarillado de Bogota), EPM (Empresas Publicas de Medellin) and Aguas de Manizales. Aguas de Manizales agreed to take over the Cartagena concession from AgBar, but this was blocked by Bogota city council. It is developing management contracts in two other regions. EPM, together with an employees' pension fund, has taken on a management contract in Bogota, and is bidding for work in Peru.

In Uruguay, a referendum decided to make water privatization illegal, resulting in the renationalization, under OSE, of the two privatized concessions. In Venezuela, the state has funded development of water services through community organizations in Caracas and peri-urban areas (Lobina and Hall, 2007).

4. Investment

Water systems require extremely high levels of investment. One of the purposes of privatization has been to obtain investments necessary to extend or improve systems without increasing government borrowing. This has been a common driver for privatization in the north – for example in the UK – and in developing countries in the south.

In the 1990s the World Bank and donor agencies promoted a strategy to develop water systems in developing countries through privatization. This was expected to deliver finance for investments, efficiency improvements, and better governance than they believed possible through the public sector in developing countries. It was expected that multinational companies would be attracted by a large new profitable market, and that the process would be welcomed by populations disillusioned with the corruption and inefficiency which the World Bank associated with the public sector. It was so central to donor policies that a World Bank official

Table 3.4 Renationalization and remunicipalization of water services in South America, 2007

Country	City/region	Former multinational	Public sector entity	New owners (%) National	New owners (%) State/region/province	New owners (%) Municipal	Employees/union
Argentina	Buenos Aires	Suez	AySA	90			10
	B A (province)	Azurix	Aguas Bonaerense SA		90		10
	Tucuman	Veolia	Sapem/OST		90		10
	Santa Fe	Suez	Aguas Santafesinas		51	39	10
Bolivia	La Paz/El Alto	Suez	Epsas	100			
	Cochabamba	Bechtel/UU	Semapa			100	
Uruguay	Maldonado	Aguas de Bilbao	OSE	100			
Venezuela	Hidrolara	SAUR	State/municipalities		50	50	

Source: Lobina and Hall 2007.

told an international conference in 2000 that 'there is no alternative' to privatization.

The private contracts have however failed to deliver significant new investment in water infrastructure in developing countries. This section examines the impact of privatization on water connections in Africa and Asia; and secondly by looking at the major cases in middle income countries in Latin America; thirdly by examining some weaknesses in the private investments in BOT dams and water treatment plants; and fourthly by comparing progress briefly with the extension of water connections in previous years.

Investment in water in the UK did increase after privatization. The next section examines how this compared with previous trends, and traces a rapid move from equity to debt by the companies, as well as the background of government assistance. The final case looks at public finance and solidarity mechanisms used to finance water extensions in southern and eastern Europe.

4.1 Africa and Asia

After 15 years, only about 600 000 households have been connected as a result of investment by private water operators in sub-Saharan Africa, South Asia, and east Asia (outside China) – representing less than 1 per cent of the people who need to be connected in those regions to meet the UN Millennium Development Goals (MDGs). One reason for this is the selectivity of the private sector, on both a macro and a micro scale. No private concessions were set up in the whole of south Asia, for example. Some contracts allowed selective provision: Suez' contract at Stutterheim, in South Africa, signed in 1993, allowed the company to 'cherry-pick' the profitable white and coloured areas, which already received dependable water supplies, while much of the official Stutterheim township (Mlungisi) and the unofficial neighbouring townships (Cenyu, Kubusie, Cenyulands) remained almost entirely outside the network. A second reason is that the great majority of contracts in Africa are lease or affermage contracts, under which the responsibility for investments in extension to the system remains with the government or municipality. In the cases of Cote d'Ivoire and Senegal, for example, which are often quoted as successes, the investment in extensions is government financed; the private companies are responsible only for maintenance of the existing system. In Senegal, public and donor finance across the 10 years of the contract totals US$230 million, while the finance provided by the private company SDE is about US$20 million over the same period.

Table 3.5 Investment under-performance by Aguas Argentinas, 1993–98 (in millions of pesos/dollars at supply values)

	1993	1994	1995	1996	1997	1998	Total
Investments committed in original bid	101.5	210.52	302.91	362.36	229.10	83.07	1289.46
Investments realized ($)	40.93	144.55	132.17	100.49	109.52	15.41	543.07
Under-performance ($)	−60.57	−65.97	−170.74	−261.87	−119.58	−67.66	−746.39
Under-performance as % of investment committed	−59.8	−31.3	−56.4	−72.3	−52.2	−81.5	−57.9

Source: Lobina and Hall, 2007.

4.2 Buenos Aires

During the period of the private water concession in Buenos Aires, which ran from 1993 to 2006, although services improved the company did not meet the targets for investment, nor for quality. The water regulator ETOSS estimated that between 1993 and 2002 Aguas Argentinas delivered only 61 per cent of the total investment due (Ducci, 2007).

These problems did not commence with the economic crisis which hit Argentina in 2001. Between 1993 and 1998 the company delivered only 42 per cent of the originally agreed investments, saving the company a total of US$746.4 million. Even after several renegotiations of the investment targets, Aguas Argentinas realized only 60 per cent of projected new connections to the water supply network and 40 per cent of projected investments in the expansion of the sewerage network (Azpiazu and Forcinito 2002).

Despite this under-performance in terms of investment, the average water bill increased 88 per cent between 1993 and 2002, compared with general inflation of only 7 per cent; the company achieved a return on assets of 21 per cent from 1994 to 2001, until the economic crisis; and used a much higher level of debt than implied in the original tender.

The need for solidarity finance was also emphasized during this concession. Poor slum areas remained unconnected to the system, whose households were unable to pay tariffs that would cover the cost of supplying them with water. The private company therefore proposed to implement a solidarity surcharge (known as the SUMA) on existing users, to pay for the cost of supplying the poor. This was however resisted by consumers who won a court case declaring the imposition of the charge illegal: and the company was authorized to collect the charge only after the intervention of the national government and the mayors of various districts in Buenos Aires.

Table 3.6 Population connected to water and sewerage services in Buenos Aires by new extensions to the system: Projected and actual, May 1993–December 1998

	Water (thousands)	Sewerage (thousands)
Connection targets		
• According to original bid	1709	924
• According to Resolution Etoss N° 81/94 (First Negotiation)	1764	925
• According to Decree N° 1,167/97 (Second negotiation)	1504	809
Actual connections constructed		
• Regularization of illegal users	172	152
• Real expansion of the network	917	399
Degree of effective compliance (Excluding regularization of illegal users)	Percentage	Percentage
With respect to the original bid	54%	43.2%
With respect to targets after second negotiation	60%	40.3%

Source: Azpiazu and Forcinito, 2002.

4.3 Chile

Water privatization in Chile is often held up as a successful case, accompanied by a high level of efficiency and significant new investment. But it is not clear that this has delivered significant economic and social gains compared with what would have happened otherwise.

The investment following the privatization did not deliver any significant extension of the system – nearly 100 per cent of households were connected to water, and around 90 per cent connected to sewerage, before privatization. The new investment was in sewerage treatment works. This implemented a government policy commitment made before privatization, through a regulatory framework that allowed the full cost of this investment to be raised from water and sewerage charges on consumers. The advantages of carrying out this investment through privatization were that the large investment does not appear on the government balance sheet, and that the government could distance itself from the price rises (Bitran and Valenzuela, 2003; Bitran and Arellano, 2005; Lobina and Hall, 2007).

Operations were already comparatively efficient: the utility in Santiago, for example, had been described by the World Bank as the most efficient utility in South America while under public ownership and management. An early review of performance following privatization found that the relative performance of public and private companies varied according to the dimension examined. Private companies increased profitability faster, increased prices faster, and increased unaccounted for water (leakage) faster than public companies; but private companies reduced labour costs faster, reduced administrative costs faster, and invested more in sewerage treatment.

The sharp increase in profitability was made possible primarily because the average price of water in Chile trebled between 1989 and 2002, which even supporters of the privatization believe was 'unlikely to survive public scrutiny if not accompanied by vigorous and sustained economic growth, which has helped make it possible for households to pay the price'. This was supported by a regulatory framework which allowed a return on assets of at least 10 per cent, and a readiness by the government to reach informal agreements with company negotiators. Some concessions were made more attractive by being of indefinite duration. The result was that the private concessions achieved returns on equity of about 14 per cent by 2005, on a par with the best performers on the Chilean stock exchange. Further downstream profits were also achieved by the company winning the concession in Santiago, Suez, which awarded the $330 million contract for the construction and operation of the largest wastewater treatment plant, La Farfana, to Degremont, Suez's own engineering subsidiary (Bitran and Valenzuela, 2003).

4.4 Colombia

In 1994 the World Bank agreed to finance investment in the water system of Cartagena, on condition that the service was privatized: an agreement was signed with the mayor on his last day in office before an election. The election was won by a candidate opposed to privatization, but the World Bank insisted that the funding would be cancelled without privatization. The agreement with Acuacar, a joint venture between with Aguas de Barcelona and the council, is a lease contract so the company has no responsibility for investment finance – as in other leases, the extension is heavily financed by the World Bank and the municipality. Acuacar is also exempt from paying any lease fee for the use of the system and, in addition to sharing dividends, Aguas de Barcelona also benefits from a management fee paid by Acuacar, which was calculated as a growing percentage of Acuacar's gross income: in the first four years of operation, this management fee was fixed at 2.94 per cent, 3.37 per cent, 3.82 per cent and 4.25 per cent respectively of gross income. The municipality also took on pension obligations worth $8 million per annum for former employees (the company had dismissed all the former 1800 employees and rehired 270 of them in order to boost operating efficiency) (Lobina and Hall, 2007).

4.5 BOTs

In recent years private companies have preferred to invest in Build-Operate-Transfer (BOT) contracts, which are used in a number of countries as a way of financing the construction of new reservoirs, water treatment plants and sewage treatment plants. The principle is similar to Private Finance Initiative schemes used in the UK and elsewhere. To provide security for investors such agreements are normally guaranteed by national governments; if the municipal water distribution authority does not pay for the water for any reason, the government promises to do so. On the strength of this guarantee of government payments, the companies can borrow money for the construction costs at low rates of interest. BOT contracts do not provide investment in the distribution system itself, and so do not extend water supply to new users, although they clearly increase the capacity of the system to provide water to consumers.

These contracts may actually create extra demands on the finances of a water distribution authority, and so reduce the money available to the distribution authorities for other purposes. There are two factors that tend to produce this result. First, the terms of the original contract are crucial in determining the level of payments for 30 years. As a result, the companies have a large incentive to engage in corruption or misrepresentation in order to increase their chances of winning a contract on favourable terms, for example by exaggerating forecasts of demand for water. Secondly, the take-or-pay agreement, underpinned by government guarantee, limits the risk taken by companies, but means that the BOT contract must be paid before the water distribution authority can use its income for any other purpose, such as investing in extending the system to the poor. The take-or-pay agreement may thus impose financial demands on the water authorities and the public, even if the price of the water turns out to be unaffordable, and even if the extra water supplied turns out to be unnecessary.

For example, in Vietnam, the Thu Duc treatment plant in Ho Chi Minh City began operating in 1999. Under the contract it sold water to the city water utility at 20 cents per cubic metre, although the price charged by the utility to consumers was only 11 cents. The balance had to be subsidized by the city council. In February 2003 the contract was ended. The bulk water supply contract of Shenyang Public Utility also ended in 2002 because demand was lower than forecast and the public water authority could not afford to pay. A BOT contract in Bogota, Colombia, was terminated after the city council calculated that the project was charging ten times too much, and that it was worth paying US$80 million to buy out the contract. The contract for the Yuvacik Dam near Izmit in Turkey stated that the water would be purchased over 15 years at an agreed price. However, both industrial users and neighbouring municipalities have refused to buy water from the plant as it is too expensive. The purchase of water was guaranteed by the Turkish government, which has thus paid over the odds for water that is too expensive for its intended customers. An enquiry in 2003 recommended the investigation for corruption of nine former ministers and the former mayor of Izmit.

Table 3.7 Observed problems with BOT contracts

Country	Project	Companies	Problems for water distributor	Public guarantees	Status
China	Chengdu	Veolia	X	X	Distressed/disputed
China	Da Chang (Shanghai)	Thames Water, Bovis	X	X	Terminated
China	Shenyang	Suez	X	X	Terminated
China	Xian	Berlinwasser (Veolia/Thames)		X	Terminated
India	Bangalore	Biwater	X	X	Cancelled
India	Sonia Vihar (Delhi)	Suez	X	X	Distressed/disputed
Vietnam	Thu Duc (HCM City)	Suez, Pilecon	X	X	Terminated
Malaysia	Salangor	Puncak Niaga	X	X	
Thailand	Pathum Thani	Thames/Bovis Karnchang	X	X	
Turkey	Yuvacik (Izmit)	Thames	X	X	Distressed/disputed
Zimbabwe	10 dams plan	Biwater	X		Cancelled

Source: Hall and Lobina, 2006.

4.6 Comparison with 1980s

Even in Latin America, where private sector investment was concentrated, the expansion of the system in the 1990s was no greater than in the 1980s. Indeed, the proportion of the population connected to water supply in the 1990s – 5 per cent – was smaller than in any of the preceding three decades.

Table 3.8 Households connected to water and sewerage by decade, Latin America

	Water		Sewerage	
	m. hab	%	m. hab	%
1960	69	33	29	14
1971	152	53	59	21
1980	236	70	95	28
1990	341	80	168	39
2000	420	85	241	49

Source: Pan American Health Organization, quoted in Jouravlev (2004).

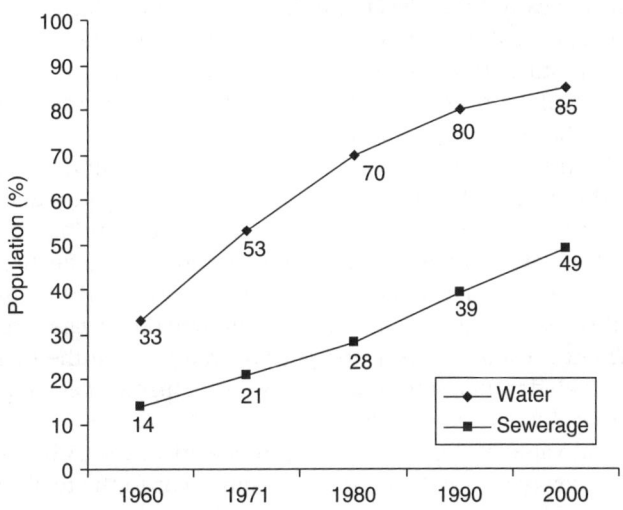

Figure 3.6 Water and sewerage connections, Latin America, 1960–2000
Source: PAHO, quoted in Jouravlev, A. (2004).

4.7 England: levels of investment

The level of actual capital investment in the water industry has been much higher since 1989 than it was in the previous decade. This is now claimed as an indicator of the success of privatization: a factsheet published by OFWAT gives the figures for investment before and after 1989, and claims: 'Under Ofwat, investment in water and sewerage services is at its highest ever level. According to OFWAT, a total of £55 billion has been invested in the 15 years since privatization, an average of £3.7 billion per year, compared with an average figure of £2 billion per year during the 1980s. This is a difference of £1.7 billion per year, or 46 per cent of all expenditure (all figures are at 2004–05 prices)' (OFWAT, 2005, 2006).

This picture however exaggerates the difference between investment levels before and after 1989. The RWAs did not make the same level of investment throughout the 1980s, but showed a clearly rising trend towards the end of the decade, recovering from the long decline in investment imposed by successive governments between 1975 and 1985. Between 1985 and 1989 investment rose steadily from about £1.6 billion to over £2.2 billion per year, so their investment had been increasing at a rate of 8 per cent per year in the second half of the 1980s. The OFWAT comparison assumes that there would have been no further increase by the RWAs, but this is unlikely: because of the legal requirements for investment (see next section) the RWAs would certainly have had to continue increasing their level of investment. Even if this increase had averaged 4 per cent per annum, half the rate they were delivering in the second half of the 1980s, they would have delivered a total investment of over £50 billion over the next 15 years: about the same as the private companies have actually achieved.

It is also true that after privatization the finance became available to pay for the necessary investment. This however was partly due to the government injecting a large amount of money, by writing off all the debts of the water companies before privatization, plus a further 'green dowry' to meet the environmental standards required by the EU. In addition to this cash injection the government allowed the private companies to make large real increases in the price of water, which the RWAs had been prevented from doing, and the private companies were freed from the limits on public sector borrowing.

The final value of the debt write-off was worth over £5 billion, and the green dowry £1.5 billion – roughly equivalent to the total received for the sale of the companies (the water and sewerage companies even gained an extra £120 million just by having these gifts in the bank in

Table 3.9 Investment level and growth rate before and after privatization in England and Wales (£billion, 2003–04 prices)

	1985	1989	2004	Average annual % growth rate
RWAs (pre-privatization)	1.6	2.2	–	8%
Privatized companies and OFWAT	–	2.2	3.6	3%

Source: OFWAT, 2006; authors' calculations.

1990/91). These public subsidies alone financed roughly one-third of all the investments in the first 10 years of privatization. There was a further subsidy in the form of tax relief on the companies' profits worth £7.7 billion. The total amount of public finance injected into the privatized water companies was thus over £14 billion (though much of the tax relief was subsequently clawed back by the 'windfall tax' introduced by the new Labour government in 1997) (OFWAT, 1995; Schönbäck et al., 2004; OFWAT/DEFRA, 2006).

When the English and Welsh water companies were privatized in 1989, the government wrote off all the existing debts, so it was entirely financed by shareholder equity; for comparison, companies in general in the UK have debts representing between 20 and 30 per cent of the total of debt and equity. The broad expectation was that as the water companies made profits investors would continue to inject new equity, and the regulator has set price caps based on assumptions about returns on capital which 'ensure that returns assumed should provide shareholders with sufficient incentives to provide additional funds, either in the form of retained earnings or new equity, to enable companies to make new investment where this is appropriate' (OFWAT/DEFRA, 2006 p. 97). But in practice the water companies preferred to finance investment at relatively low rates of interest, which allowed them to generate a higher surplus for distribution to shareholders. As a result, the gearing of the water companies has risen from zero to an average of over 60 per cent in 2006. This process has been accelerated because the regulator has overestimated the cost of debt in setting prices, so that price-caps allow the companies to charge users: 'at a level significantly higher than the actual cost of debt over the period. As a result customers/users have paid higher prices and returns on equity have been higher than expected when the price control was set' (OFWAT/OFGEM 2006; CEPA 2007 p. i).

Instead of shareholders putting money into the industry there has been a significant *withdrawal* of shareholder equity from the water companies (OFWAT/DEFRA, 2006). Different methods of withdrawing equity were adopted. The most extreme version took place in Wales, where the private water and energy utility was taken over by a consortium of USA energy companies, who wanted to abandon the water business altogether. They transferred all the assets, liabilities and statutory functions to a not-for-profit company, run by an appointed and self-perpetuating group of individuals, and financed entirely by debt. (This entity is neither elected by citizens nor owned by shareholders or customers, but is often wrongly described as a cooperative or a mutual.) Another company proposed complete withdrawal of equity from Yorkshire Water by selling the company to a consumer cooperative, but this was abandoned as a result of fierce local opposition.

Other companies have simply reduced their equity stakes and replaced them with debts, including Anglian and Southern water. The water only companies have undergone a number of similar restructurings: for example, East Surrey issued a £100 million bond; Mid Kent Water was purchased by a management buyout, the Swan Group, funded predominantly by debt from WestLB; there was a similar deal at Portsmouth Water, backed originally by Royal Bank of Scotland; Veolia's former shareholdings in Bristol Water and South Staffordshire were purchased by an investment fund, Ecofin Water and Power. This represents a return to the same form of finance used by public sector water operators – indeed, a significant part of the borrowing has been from the European Investment Bank (EIB), a public sector bank owned by the European Union which is able to lend at very good rates.

4.8 Solidarity finance in southern Europe and European countries in transition

The traditional use of solidarity finance has continued in Europe. The European Union collects taxes from all EU Member States and distributes them through its cohesion policy. On average the EU collects about €20 in taxes from every person in the EU each year to support investment in water and sanitation alone. During the period 1994 to 1999, environmental investment financed from the Structural Funds amounted to over €9 billion (ENEA, 2006; European Commission, 2006).

The impact on coverage in less wealthy regions and Member States was significant:

> In Greece, the number of urban areas connected to main drainage almost doubled between 1993 and 1999, increasing the population covered to over 70 per cent. In Ireland, the proportion covered rose from 44 per cent in 1991 to 80 per cent in 1999. In Portugal, the population connected to drinkable water supply rose from 61 per cent in 1989 to 95 per cent in 1999 and that connected to main drainage from 55 per cent in 1990 to 90 per cent in 1999. The Funds also helped to increase water supply in regions with a serious shortage. In Italy, for example, supply was expanded by over a third over the programming period. (European Commission, 2006)

Overall, this central support for infrastructure and other measures had a major effect on economic growth; in Greece, GDP in 1999 was 9.9 per cent higher than it would have been without the central cohesion funds, in Portugal 8.5 per cent higher (European Commission, 2006).

The countries of central and eastern Europe provide another interesting study of the relevance of public finance mechanisms for investment in water services. After the collapse of the communist regimes around 1990, their water services were restructured, mainly through decentralization of responsibility to municipalities. The private water companies of France and the UK took the opportunity to obtain concessions in a number of cities, with particular success in the Czech Republic and Hungary; in Poland, however, with the exception of Gdansk, the water services remained in municipal hands.

The systems already provided nearly universal connection, and so relatively little investment was required for extension of the system, but suffered from varying degrees of poor maintenance and technical inadequacy, and a general need for investment in wastewater treatment. These needs became magnified as the countries joined the EU because of the requirements of EU water and environmental standards.

The financing of this investment came from four major public sources. First, for the countries bordering the Baltic Sea – Poland and the Baltic republics – international finance was made available through grants and development bank loans associated with the Helsinki Commission, dedicated to improving the water quality of the Baltic. Wastewater treatment plants, in particular, were financed in this way. Secondly, the EU's own development banks – the European Investment Bank (EIB) and the European Bank for Reconstruction and Development (EBRD) – made loans for

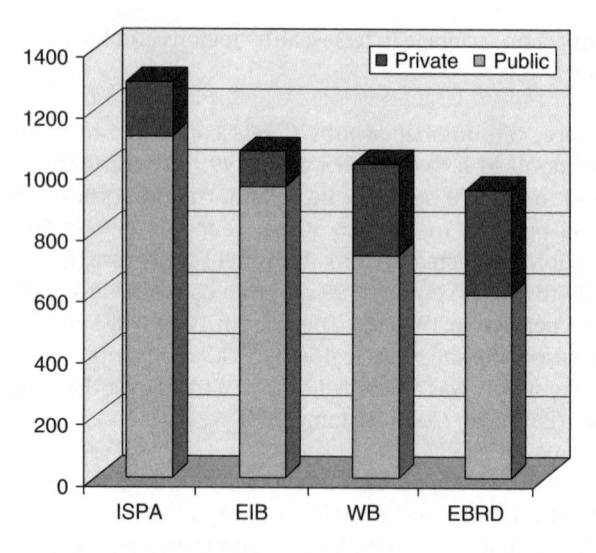

Figure 3.7 Development bank finances for public and private water projects in central and eastern Europe 1990–2002
Source: Hall and Lobina, 2007b.

investment in the water sector. These loans were made to both public and privatized water operations, with the EBRD in particular making loans to a number of municipalities in Poland and the Baltic states without requiring government guarantees. Thirdly, as the countries prepared for EU membership they became eligible for large grants from the cohesion and infrastructure funds of the EU itself. Fourthly, national and local governments themselves used tax revenues to finance water investment. In Poland, for example, municipal or enterprise funds financed around 45–50 per cent of capital investment in environmental improvements in the 1990s, with the balance coming largely from environmental funds and domestic loans. In Hungary, central government continues to finance most capital investment in the water system, even in cities where this is privatized (De la Motte, 2007; Hall and Lobina, 2007b).

Finally, the national and local governments of the countries also continued to provide substantial finance for investment, even in cities where the operation had been privatized. In Poland, for example, water operations were supported through both municipal funds and a national environmental fund, financed from taxation; in Hungary, the cost of investments in the system is still largely borne by national government funds (De la Motte, 2007; Boda *et al.*, 2006).

5. Prices

Water privatization is often associated with increases in prices and the opposition to privatization often centres around these price rises. Prices may be affected by a number of external and general factors, including the relative cost of collecting and distributing water, and also by trends in investment.

Because water operators have a monopoly of an essential service, it is expected that private water companies will have a constant incentive to try and extract monopoly profits by excessive pricing (and underinvestment). This section examines evidence from France and the UK which supports this view, despite the presence of experienced or professional regulation.

5.1 France

It is possible to compare the price of water charged by public and private water operators in France, where municipalities still provide water directly in a quarter of cases. Data collected annually shows consistently that the prices charged by private companies are significantly higher than the prices of municipal operators. In 2004 the average price charged by private companies was 29 per cent higher than that charged by municipal companies (IFEN, 2007).

This difference could be due to other factors affecting the cost of water supply, such as the requirement for treatment, the density of the network, and the condition of the network. However, a recent analysis of comprehensive data covering 5000 municipalities in France controlled for all such factors, and found that privatization alone accounts for a difference of 17 per cent in prices:

> choosing any kind of PPP [i.e. private water companies] over direct public management seems to increase the average retail price of water in a municipality....the average price for delivery of 120 cubic meters of water in a year jumps from approximately €151 to €176 when a public authority chooses a lease contract instead of managing its own water distribution

and concluded that 'consumers pay more when municipalities choose PPPs' (Chong *et al.*, 2006 pp. 163, 150). As the authors note, this result may be the result of collusion strategies and/or corruption, as revealed in a public audit report in 1997 (Cour des Comptes 1997).

Water Price in France (€/m³)

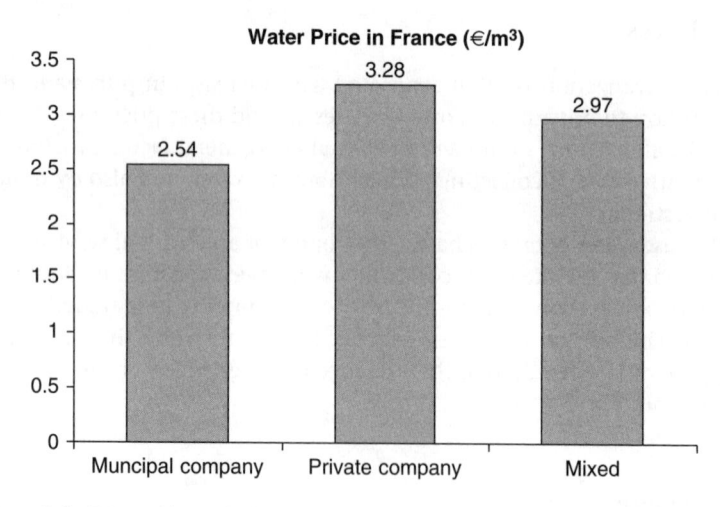

- Muncipal company: 2.54
- Private company: 3.28
- Mixed: 2.97

Figure 3.8 Price of household water in France, 2004
Source: IFEN, 2007.

The poor suffer all the impact of this difference. A study of affordability of water in France found that the poorest households receive significantly higher water bills under private operators, especially in concession contracts, whereas rich households receive slightly smaller bills. By contrast, the same study supports the view that politics also made a difference to prices, with left-wing parties associated with more affordable water: 'A high proportion of votes to the socialist or to the communist party at the last local election results in a lower share of income spent on water charges' (Reynaud, 2006, p. 20).

5.2 UK: prices and gaming under regulation

The evidence from England and Wales is also consistent with this view that privatization itself tends to generate higher prices as a result of concerted activity by the companies. In cash terms, the average annual bill for water and sewerage rose from £120 per year in 1989 – the year of privatization – to £294 in 2006, an increase of 245 per cent in 17 years. In real terms, it represents a rise of 39 per cent over and above the general rate of inflation. A breakdown of the component elements in the water bills shows that operating costs have remained roughly constant in real terms: the increase in customers' bills is almost entirely due to the various elements associated with the capital – capital charges, interest

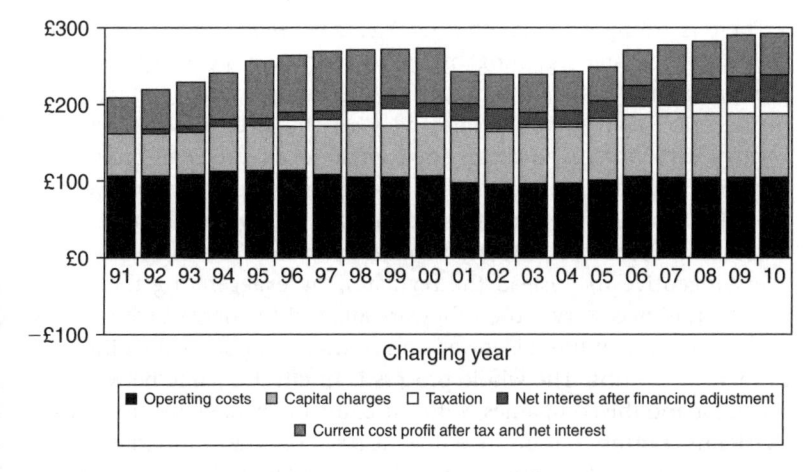

[Chart showing components of average household bill, £0 to £300, years 91 to 10, with legend: Operating costs, Capital charges, Taxation, Net interest after financing adjustment, Current cost profit after tax and net interest]

Figure 3.9 Components of the average household bill in England and Wales 1991–2004 (constant prices)
Source: OFWAT/DEFRA, 2006.

and profits – which have nearly doubled, in real terms, over this period (OFWAT/DEFRA, 2006).

Some such increase would be expected because of the increased investment resulting from the requirements of EU directives, but international comparisons of changes in volumetric charges to industrial users indicate that the UK privatization has had a distinct additional effect. In 1988–89, the last year of public ownership, the NUS survey of water costs for industrial users showed that British companies paid relatively low charges for water: Britain was ninth in the NUS league table, behind five of its EC counterparts including Italy. The NUS figures show that in 1988–89, the cost of water in the UK was less than half that in Australia and West Germany. In 2005 the corresponding NUS survey showed the UK was in third highest position, with costs nearly double those of Australia, 70 per cent higher than in Italy, and only 18 per cent lower than Germany (NUS, 2005).

Under the privatized system, the regulator OFWAT is responsible for setting price limits and incentives so that the companies, while making a profit, can deliver the service, and the prices, that consumers want. The water companies are responsible to their shareholders for achieving the best possible return. The system is intended to result in regulations which create incentives for the companies to improve their performance, but also creates incentives for the companies to try and arrive at a more favourable deal for themselves at the expense of consumers.

There is strong evidence that OFWAT has been unable to deal with active and persistent 'gaming' by the companies in order to gain higher profit margins. This gaming happens around the price caps set by OFWAT in the price reviews, which effectively set the level of water prices in England five years in advance. The companies submit their projections of expenditure and claim that they need to increase prices to cover this spending. OFWAT then has to try and make its own assessment of the accuracy of these forecasts, and then set the prices. The companies have every incentive to mislead the regulator, by exaggerating the capital expenditure necessary – then they are allowed to charge higher prices, but the real expenditure is lower, and so they can pocket the difference as increased profit. The whole process is in effect a game between the regulator and the companies, with the company behaviour summarized by Helm: 'a utility has an incentive at price-setting to inflate the asset base, to inflate CAPEX, and to argue for a high OPEX. It will also want to maximize the assumed cost of capital. The higher the expected costs, the higher the added value to shareholders' (Helm, 2003; see also Wietze and Bakker, 2005). The extent of it was indicated by a series of disclosures and confessions of systematic attempts to mislead the regulator. A manager of Severn Trent, David Donnelly, said in 2004 that he had been instructed by his bosses to exaggerate figures of debts owed by non-paying customers: Severn Trent was charged with fraud at the end of 2007. Other companies confessed to similar 'errors': Southern Water admitted mistakes about its responses to customers, and failure to make payments due to customers; Thames Water and Severn Trent itself admitted that they had misrepresented data on its response to customer enquiries, which also affects customer bills; Tendring Hundred admitted it had made an 'accounting error' in its estimates of income from metered customers, and overcharged customers £5 per head as a result of this unfortunate mistake (OFWAT, 2006a).

6. Efficiency

The question of comparative efficiency is central to the arguments over the economic merits of privatization, in water as in other sectors. This section examines the empirical evidence on this issue, both in general and specifically in the water sector. A paper by the IMF in 2004 noted that the issue is crucial for justifying any form of PPP because public sector borrowing is invariably cheaper than private sector borrowing, and so the key question is whether PPPs result in efficiency gains that more than offset the higher borrowing costs. The IMF paper warns against making a priori assumptions of superior private sector efficiency: 'Much of the case for PPPs rests on the relative efficiency of the private sector. While there is an extensive literature on this subject, the theory is ambiguous and the empirical evidence is mixed.... It cannot be taken for granted that PPPs are more efficient than public investment and government supply of services' (IMF, 2004, para 25).

Contrary to widespread assumptions, and the conclusions of Megginson and Netter (2001) in favour of private ownership, the overall results of empirical studies are inconclusive, and do not support any general conclusion of superior private sector efficiency. In monopolies, typified by the water sector, there are also theoretical arguments for expecting better performance from public sector companies (Willner and Parker, 2007; Willner, 2001). Studies of the UK privatizations support this. A review in the late 1990s concluded that there is 'little evidence that privatization has caused a significant improvement in performance. Generally the great expectations for privatization evident in ministerial speeches have not been borne out' (Martin and Parker, 1997). A comprehensive review in 2004 was 'unable to find... evidence that output, labour, capital and TFP productivity in the UK increased substantially as a consequence of ownership change at privatization compared to the long-term trend' (Florio, 2004).

6.1 Comparative public–private efficiency in the water sector

In the water sector there is now a considerable body of empirical evidence supporting the view that private operators are not intrinsically likely to be more efficient than public sector operators. A World Bank paper in 2005 summarized the econometric evidence thus:

> Probably the most important lesson is that the econometric evidence on the relevance of ownership suggests that in general, there is no statistically significant difference between the efficiency performance of public and private operators in this sector.... For utilities, it seems that in general ownership often does not matter as much as sometimes argued. Most cross-country papers on utilities find no statistically significant difference in efficiency scores between public and private providers. (Estache, *et al.*, 2005)

The evidence covers both developed and developing countries. The results put in perspective observations of improvements following privatizations, which assume that any improvements observed are due to private ownership, without making any comparison with the control group of public sector operators.

In Africa a 2004 study by Kirkpatrick *et al.*, covering 110 African water utilities, including 14 private, found no significant difference between public and private operators in terms of cost. A much smaller earlier study by Estache and Kouassi of water operators in Africa in 2002 did find that private operators were more efficient, but only included two private operators, and institutional quality was a more important factor than private ownership in explaining differences in efficiency (Kirkpatrick *et al.*, 2006; Estache and Kouassi, 2002).

A 2004 study of about 4000 sanitation operations in Brazil found that there is no significant difference between public and private operators in terms of the total variation in productivity; a further study in Brazil, published in 2007, also concluded that 'that there is no evidence that private firms and public firms are significantly different in terms of efficiency measurements' (Seroa da Motta and Moreira, 2004; da Silva e Souza *et al.*, 2007). A study of water utilities in Chile found that private operators had increased investment and labour productivity by more than public companies: though they had also increased their rates by more, and had performed worse in dealing with unaccounted for water (Bitrán and Valenzuela, 2003).

A paper published by the Brookings Institute in 2004 studied the growth in water and sanitation connections in cities in Argentina, Bolivia and Brazil, both in cities that had private sector participation and in cities that had no private sector involvement. Using household level data, it is the most comprehensive comparative survey of connections under private and public management – other case studies have focused on private sector operations alone and assumed that any improvements observed were due to private ownership. It concluded that 'while connections appear to have generally increased following privatization, the increases appear to be about the same as in cities that retained public ownership of their water systems' (Clarke *et al.*, 2004).

In Asia, a similar mixed picture emerges. In 2004 the Asian Development Bank conducted a survey of 18 cities in Asia, which included two cities with private sector concessions – Manila and Jakarta. These were performing significantly worse than most public sector operators on four indicators of coverage, investment, and leakage: on six indicators (unit production costs, percentage of expenses covered by revenue, cost to consumers of constant level of usage per month, 24-hour supply, tariff level, connection fee) their performance is middling, not outstanding; the private cities perform relatively well on two indicators: revenue

Table 3.10 Selected ADB water indicators for 18 Asian cities

		Manila (private)	Jakarta (private)	Average of 18 cities (public)
Water coverage	(%)	58	51	79
Sewerage access	(%)	7	2	51
Non-revenue water	(%)	62	51	34
Capital expend/connection	(US$)	18	47	88

Source: ADB, 2004.

collection efficiency and minimizing the number of staff per 1000 connections (ADB, 2004). An earlier study on 50 cities in Asia also concluded that 'The results show that efficiency is not significantly different in private companies than in public ones' (Estache and Rossi, 2002). A study of towns in Cambodia found that consumer satisfaction and service continuity was higher (however, prices were higher and not affordable for all), although the privatized towns had been selected by the operators and so may have been better performing anyway (Garn et al., 2002).

The picture is similar in respect of operators in OECD countries. A Brookings Institute paper in 2005 looked at public and private water operators in the USA in terms of regulatory compliance and household expenditure on water. It found that 'when controlling for water source, location fixed effects, county income, urbanization, and year, there is little difference between public and private systems' (Wallsten and Kosec, 2005).

Comparisons over time in the UK suggest that efficiency of the water sector, measured by productivity, has not improved since privatization, and may actually have got worse. An analysis of productivity growth in the five years before privatization, and the 10 years after privatization, concluded that: 'despite reductions in labour usage, total factor productivity growth has not improved since privatization' (Saal and Parker, 2001). A further study using a different methodology showed that total factor productivity may have improved after 1995 but 'neither paper finds any evidence of an increase in TFP growth that can be directly attributed to privatization' (Saal, 2003). A third study, with a further change in methodology, concluded that productivity had declined, showing

> that while technical change improved after privatization, productivity growth did not improve, and this was attributable to efficiency

losses as firms appear to have struggled to keep up with technical advances after privatization.... average efficiency levels were actually moderately lower in 2000 than they had been at privatization. (Saal et al., 2007 pp. 127, 138)

Despite the technical advances the private water companies are not spending more on research and development (R&D) than before privatization: 'many companies' research and developments budgets have all but disappeared' (House of Lords, 2006). R&D has very high returns, and even higher social returns, but is risky and the benefits may not be limited to the company that does the research. As a result: 'Private markets, including competitive markets, are expected systematically to under-provide R&D in relation to what is socially desirable' (Thomas, 2004, p.6; Rosenberg, 1990).

This sector-specific result mirrors the results of studies of the UK privatizations in general, which have concluded that there is 'little evidence that privatization has caused a significant improvement in performance. Generally the great expectations for privatization evident in ministerial speeches have not been borne out', and were 'unable to find ... evidence that output, labour, capital and TFP productivity in the UK increased substantially as a consequence of ownership change at privatization compared to the long-term trend' (Martin and Parker, 1997; Florio, 2004).

It is finally worth noting the evidence of case studies especially in Africa, where repeated outbreaks of water-related diseases like cholera and typhoid in areas run by private water companies have often been accompanied by reports of extremely ineffective management responses, for example in time taken to repair systems. This evidence does not establish a comparative inferiority, but does provide further support for the view that superior performance should not be consistently expected from the private sector (Hall and Lobina, 2006).

7. Conclusions

The experience of water privatization over the 20 years since the late 1980s can be said to have confirmed the experience of a century earlier. Private companies can operate and invest in a water system based on market principles where customers are connected according to their willingness and ability to pay. Such markets exist now, both in the shelves of expensive bottled water in supermarkets and in the vendors of water in slums with no piped connections. The development of a universal piped water service, however, places demands on private investors that are likely to be resisted – as it was a century ago in Europe and the USA – as marginal customers are unable to pay enough to make the connection profitable. The lack of private investment in developing countries is simply because it would be bad business, as the private companies themselves explained to the World Bank.

In developing countries where extensions are imperative for economic, social and health reasons, public finance is now – as in the nineteenth century – the only reliable mechanism for delivering the extensions. The key element in public finance is the ability to use legitimate solidarity mechanisms to redistribute the cost of financing extensions from the unconnected to the society and economy as a whole.

In effect, in the last 20 years, governments and municipalities in developing countries have found ways of injecting public finance – using national or donor sources, or development banks – as a way of sustaining investment. Extensions of water services have happened, and public finance has been the mechanism for doing this, even where water systems have been privatized. The use of lease contracts, rather than concessions, in Africa, is one way of operating the system while investing in this way; the use of direct public sector operations is another – mirroring the mix of policies developed by French municipalities a century ago.

In systems that are already virtually complete, as in the north, private companies still suffer from the disadvantage of a higher cost of capital. The experience of the UK private sector shift from equity to debt implies the loss of the distinctive incentive of equity finance, in favour of debt financing which is cheaper through the public sector.

These disadvantages in terms of capital finance have traditionally been regarded as offset by efficiency gains, but the empirical evidence heavily supports the presumption of no significant difference in efficiency.

The evidence from France and England further supports the presumption that private companies can and will find ways of driving up prices, and/or under-investing, to obtain monopoly profits, including corruption. In the absence of efficiency gains, and in the absence of any advantage in financing investments, there is no obvious advantage held by the private sector to offset this risk of monopoly behaviour and the transaction costs of attempting to control it. The fact that even lease and management contracts have been terminated supports the view that these effects are significant.

There may also be significant institutional and social gains to a country and a region from using its own expertise rather than outsourcing such a function, as well as a reduction in the opportunity for corruption.

The continued greater popularity of direct public sector operations thus echoes the popularity of the public sector option across Europe and north America a century earlier.

Water privatization retains two great attractions, arising from institutional factors. One is that it can be used to disguise investment for public policy purposes as private investment. This is of advantage in any situation where there are constraints on public authority spending, borrowing and debt. The second is that international development banks continue to set privatization as a lending condition. It remains to be seen how long these institutional factors can continue to impose policies that are at odds with the lessons of both recent and historical experience.

Note

1. This article draws on research carried out for PSI, EPSU, WDM and others, and many reports published by PSIRU on its website at www.psiru.org.

References

ADB (2004) 'Water in Asian Cities – Utilities Performance and Civil Society Views', Asian Development Bank. January. http://www.adb.org/Documents/Books/Water_for_All_Series/Water_Asian_Cities/regional_profiles.pdf

Azpiazu, D. and Forcinito, K. (2002) 'Privatization of the water and sanitation systems in the Buenos Aires Metropolitan Area: regulatory discontinuity, corporate non-performance, extraordinary profits and distributive inequality', paper presented at the First PRINWASS Project Workshop, University of Oxford, 22–23 April, 2002. (http://www.geog.ox.ac.uk/~prinwass/Azpiazu_Forcinito.PDF)

Barraqué, B. (1995) 'Les politiques de l'eau', Paris: Éditions La Déconverte.

Bitrán, G. and Valenzuela, E. (2003) 'Water services in Chile: Comparing private and public performance', *Public Policy for the Private Sector*. No. 255, March. http://rru.worldbank.org/PapersLinks/Open.aspx?id=1998

Bitrán, G. and Arellano, P. (2005) 'Sending the right signals to utilities in Chile', *Public Policy for the Private Sector*, No. 286 March.

Blokland, M., Braadbaart, O. and Schwartz, K. (eds) (2000) *Private Business, Public Owners – Government Shareholdings in Water Enterprises*. Published for the Ministry of Housing, Spatial Planning and the Environment of the Netherlands.

Boda, Z., Scheiring, G., Hall, D. and Lobina, E. (2006) 'Social policy, regulation and private sector water supply: the case of Hungary' *UNRISD Working Paper*. Forthcoming

Castro, J.E. (2004) PRINWASS Project: Final Report 2004. http://users.ox.ac.uk/~prinwass/PDFs/PRINWASS%20D33.zip

CEPA (2007) 'Indexing the allowed rate of return. Final report for ORR/OFWAT', Sept. http://www.ofwat.gov.uk/aptrix/ofwat/publish.nsf/AttachmentsByTitle/pr0903_cepareport.pdf/$FILE/pr0903_cepareport.pdf

Chong, E., Huet, F., Saussier, S. and Steiner, F. (2006) 'Public–Private Partnerships and prices: Evidence from water distribution in France', *Review of Industrial Organization*, 29: 149–69.

Clarke, G., Kosec, K, and Wallsten, S. (2004) 'Has private participation in water and sewerage improved coverage?: empirical evidence from Latin America', *Working paper* 04-02 AEI-Brookings Joint Centre for Regulatory Studies, January http://www.aei-brookings.com/admin/authorpdfs/page.php?id=325

Cour des Comptes (1997) 'La gestion des services publics locaux d'eau et d'assainissement', *Les éditions du Journal*, Paris.

Cutler, D. and Miller, G. (2005) 'Water, water everywhere: Municipal finance and water supply in American cities' David Cutler and Grant Miller, March http://www.economics.harvard.edu/faculty/dcutler/papers/cutler_miller_C&R_3_16_05.pdf

da Silva e Souza, G., Coelho de Faria, R. and Moreira, T. (2007), 'Estimating the relative efficiency of Brazilian publicly and privately owned water utilities: A stochastic cost frontier approach', *Journal of the American Water Resources Association*, 43(5): 1237–44.

De La Motte (2007) 'A tale of two cities: Public participation and sustainability in decision-making on water systems in two Polish cities', *Utilities Policy*, 15(2).

Ducci, J. (2007) 'Salida de operadores privados internacionales de agna en América Latina', Washington D.C.: Inter-American Bank (http://www.iadb.org/sds/doc/SOPIAALWEB.pdf)

ENEA (2006) *Making the Structural and Cohesion Funds Water-Positive*. European Network of Environmental Authorities (ENEA), February. http://ec.europa.eu/environment/integration/pdf/final_handbook.pdf

Estache, A. and Kouassi, E. (2002) 'Sector organization, governance, and the inefficiency of African water utilities', *World Bank Research Working Paper* 2890. http://rru.worldbank.org/Documents/PapersLinks/1453.pdf

Estache, A. and Pinglo, M.E. (2004) 'Are returns to private infrastructure in developing countries consistent with risks since the Asian crisis?', *World Bank Policy Research Working Paper* 3373, August. http://www-wds.worldbank.org/external/default/WDSContentServer/IW3P/IB/2004/09/09/000009486_20040909120804/Rendered/PDF/wps3373Asia.pdf

Estache, A. and Rossi, M. (2002) 'How different is the efficiency of public and private water companies in Asia?', *World Bank Econ Review* 16: 139–48.

Estache, A., Perelman, S. and Trujillo, L. (2005) 'Infrastructure performance and reform in developing and transition economies: Evidence from a survey of productivity measures', *World Bank Policy Research Working Paper* 3514, February. http://wdsbeta.worldbank.org/external/default/WDSContentServer/IW3P/IB/2005/03/ 06/000090341_20050306101429/Rendered/PDF/wps3514.pdf

Euromarket (2004) 'Analysis of the strategies of the water supply and sanitation operators in Europe'. http://www2.epfl.ch/webdav/site/mir/shared/import/migration/D3_Final_Report.pdf Accessed 17 January.

European Commission (2006) 'The contribution of structural policies to economic and social cohesion: results and prospects', European Commission. http://ec.europa.eu/regional_policy/sources/docoffic/official/reports/p3221_en.htm

Florio, M. (2004) *The Great Divestiture*. MIT Press.

Galishoff, S. (1980) 'Triumph and failure: the American response to the urban water supply problem 1860–1923', in M.V. Melosi (ed.) *Pollution and Reform in American Cities 1870–1930*, Austin, TX and London: University of Texas Press, pp. 35–57.

Gandy, M. (2004) 'Rethinking urban metabolism: Water, space and the modern city', *City* 8(3) December.

Garn, M., Isham, J. and Kahkonen, S. (2002) 'Should we bet on private or public water utilities in Cambodia? Evidence on incentives and performance from seven provincial towns', *Middlebury College Working Paper Series* 0219. http://ideas.repec.org/p/mdl/mdlpap/0219.html

Hall, D. (1999) 'Privatization, multinationals, and corruption', *Development in Practice*, 9(5), November.

Hall, D. and Lobina, E. (1999) 'Trends in the water industry in the EU', 1999. http://www.psiru.org/reports/9902-W-Eur.doc

Hall, D., Lobina, E. and De La Motte, R. (2003) 'Water privatization and restructuring in Central and Eastern Europe and NIS countries, 2002', PSIRU. http://www.psiru.org/reports/2003-03-W-CEENIS.doc

Hall, D. and De La Motte, R. (2004) 'Making water privatization illegal: New laws in Netherlands and Uruguay', November, PSIRU University of Greenwich. http://www.psiru.org/reports/2004-11-W-crim.doc

Hall, D., Lanz, K., Lobina, E. and De La Motte, R. (2004) 'International Context (revised June 2004). Watertime project', www.watertime.net/reports/Docs/WP1/D7_Int_Context_final-revb.doc Accessed 1 February.

Hall, D., Lobina, E. and De La Motte, R. (2005) 'Public resistance to privatization in water and energy', *Development in Practice*, 15(3&4).

Hall, D. and Lobina, E. (2006) 'Pipe Dreams' WDM/PSI/PSIRU University of Greenwich. http://www.psiru.org/reports/2006-03-W-investment.pdf

Hall, D. and Lobina, E. (2007a), 'Water as a public service', PSIRU University of Greenwich January. http://www.psiru.org/reports/2007-01-W-waaps.pdf

Hall, D. and Lobina, E. (2007b), 'International actors and multinational water company strategies in Europe, 1990–2003', *Utilities Policy* 15(2): 64–77.

Hall, D. and Lobina, E. (2008) 'From a private past to a public future? The problems of water in England and Wales', PSIRU University of Greenwich, February. http://www.psiru.org/reports/2008-02-W-UK.doc

Hellman, J., Jones, G. and Kaufmann, D. (2003) 'Seize the state, seize the day: State capture and influence in transition economies', *Journal of Comparative Economics*, 31: 751–73.

Helm, D. (2003) 'British utility regulation: theory, practice and reform', *Oxford Review of Economic Policy*, 10(3).

House of Lords (2006) *House of Lords Science And Technology Committee: Water Management* May. http://www.parliament.uk/hlscience/

IFEN (2007) 'La facture d'eau domestique en 2004', http://www.ifen.fr/uploads/media/de117.pdf

IMF (2004) 'Public–Private Partnerships', International Monetary Fund, 12 March. http://www.imf.org/external/np/fad/2004/pifp/eng/031204.htm

Jouravlev, A. (2004) 'Drinking water supply and sanitation services on the threshold of the XXI century', SERIE Recursos Naturales e Infraestructura N° 74. Santiago, Chile: United Nations http://www.eclac.cl/publicaciones/xml/9/19539/lcl2169i.pdf

Juuti, P. and Katko, T. (eds) (2005), 'Water, Time and European Cities: History Matters for the Futures 2005', http://www.watertime.net/Docs/WP3/WTEC.pdf

Kaufmann, D. and Vicente, P. (2005) 'Legal Corruption' Second Draft, October. World Bank Institute. http://siteresources.worldbank.org/INTWBIGOVANTCOR/Resources/Legal_Corruption.pdf, Accessed 17 January 2007.

Kirkpatrick, C., Parker, D. and Zhang, Y. (2006) 'An empirical analysis of state and private-sector provision of water services in Africa', *World Bank Economic Review* 20(1):143–63, DOI:10.1093/wber/lhj001.

Lobina, E. and Hall, D. (2003) 'Problems with private water concessions: a review of experience', PSIRU. http://www.psiru.org/reports/2003-06-W-over.doc

Lobina, E. and Hall, D. (2007) 'Water privatization and restructuring in Latin America, 2007', September. http://www.psiru.org/reports/2007-09-W-Latam.doc

Martin, S. and Parker, D. (1997) *The Impact of Privatization Ownership and Corporate Performance in the UK*, Routledge, London.

Megginson, W.L., and Netter, J.M. (2001) 'From state to market: A survey of empirical studies on privatization', *Journal of Economic Literature* 39(2): 321–89.

Melosi, M. (2000), *The Sanitary City. Urban Infrastructure in America from the Colonial Times to the Present*. John Hopkins University Press: Baltimore.

Mycoo, M. (2005) 'Shifting Paradigms in Water Provisioning Policies: A Trinidad Case Study', *International Journal of Water Resources Development*, 21(3): 509–23.

Nilsson, D. (2006) 'A heritage of unsustainability? Reviewing the origin of the large-scale water and sanitation system in Kampala', *Uganda Environment & Urbanization*, 18(2), 2006.

NUS (2005), *2004–2005 International Water Report & Cost Survey*. NUS Consulting Group. July 2005

OFWAT (1995) *Financial performance and capital investment 1994–95*.

OFWAT (2005) *Financial performance and expenditure of the water companies in England and Wales 2004-2005 report*, September. http://www.ofwat.gov.uk/aptrix/ofwat/publish.nsf/AttachmentsByTitle/fpe_report2004-05.pdf/$FILE/fpe_report2004-05.pdf.

OFWAT (2006) *Water and regulation: Facts and Figures*. http://www.ofwat.gov.uk/aptrix/ofwat/publish.nsf/AttachmentsByTitle/water_regfacts_figs.pdf/$FILE/water_regfacts_figs.pdf

OFWAT (2006b) Press notices PN 16/06, 21/06, PN26/06. www.ofwat.gov.uk

OFWAT/DEFRA (2006) *The Development of The Water Industry In England And Wales*. http://www.ofwat.gov.uk/aptrix/ofwat/publish.nsf/AttachmentsByTitle/development_of_water_industry270106.pdf/$FILE/development_of_water_industry270106.pdf

OFWAT/OFGEM (2006) 'Financing networks: A discussion paper', 2006. http://www.ofwat.gov.uk/aptrix/ofwat/publish.nsf/Content/FinancingNetworks080206

OFWAT (2007) Instrument of Appointment by the Secretary of State for the Environment of Anglian Water Services Limited as a water and sewerage undertaker under the Water Act 1989, http://www.ofwat.gov.uk/aptrix/ofwat/publish.nsf/AttachmentsByTitle/lic_anglian.pdf/$FILE/lic_anglian.pdf. Accessed 1 February, 2007.

Pezon, C. (2007) 'The role of users' cases in drinking water services development and regulation in France: An historical perspective', *Utilities Policy*, 15(2).

Pezon, C. (2006) 'The French Public Private Partnership model: genesis and key factors of success' In: UNESCO 2006 Urban Water Conflicts. UNESCO Water Science Division, Technical Documents in Hydrology. http://unesdoc.unesco.org/images/0014/001490/149032E.pdf. Accessed 17 January.

Pézon, C. (2003) 'Water supply regulation in France from 1848 to 2001: A jurisprudence based analysis', Annual conference of ISNIE 2003.

Reynaud, A. (2006), 'Private participation, public regulation and water affordability for low-income French households', September. http://www.toulouse.inra.fr/lerna/cahiers2006/06.09.202.pdf

Rosenberg, N. (1990) 'Why do firms do basic R&D (with their own money)?', *Research Policy*, 19(2): 165–74.

Saal, D. (2003) 'The Impact of Privatization on the English and Welsh Water and Sewerage Industry', Paper presented at CESifo Conference on Privatization Experiences in the EU, Munich 10-11 January. http://www.cesifo-group.de/pls/portal/docs/PAGE/IFOCONTENT/BISHERIGESEITEN/CESIFO_INHALTE/EVENTS/CONFERENCES/PRI03/PAPERS/PRI03-SAAL.PDF

Saal, D. and Parker, D. (2001), 'Productivity and price performance in the privatized water and sewage companies of England and Wales', *Journal of Regulatory Economics*, 61–90. http://www.springerlink.com/content/m3j6018112134q78/

Saal, D., Parker, D. and Weyman-Jones, T. (2007) 'Determining the contribution of technical change, efficiency change and scale change to productivity growth in the privatized English and Welsh water and sewerage industry: 1985–2000', *Journal of Production Analysis*, 28: 127–39, DOI 10.1007/s11123-007-0040-z.

Schönbäck, W., Oppolzer, G., Kraemer, R., Hansen, W. and Herbke, N. (2004) *International Comparison of Water Sectors Comparison of Systems against a Background of European and Economic Policy*. Vol. 6, English Edition Vienna, October. Table 4.36, p. 70. http://www.arbeiterkammer.at/pictures/d21/153_Band6.pdf

Seroa da Motta, R. and Moreira, A. (2004) 'Efficiency and Regulation in the Sanitation Sector in Brazil', *IPEA Discussion Paper* No. 1059.

Talbot, J. (2002), 'Is the Water Business Really a Business?' Presentation by J.F. Talbot, CEO SAUR International, World Bank Water and Sanitation Lecture Series, 13 February. http://www.worldbank.org/wbi/B-SPAN/docs/SAUR.pdf

Thomas, S. (2004) 'The Future of Research and Development in the UK Gas Industry', March. http://www.psiru.org/reports/2004-04-E-UK-Advantica.doc

Wallsten, S. and Kosec, K. (2005) 'Public or private drinking water? The effects of ownership and benchmark competition on U.S. water system regulatory compliance and household water expenditures', *Working Paper* 05–05, March. http://www.aei-brookings.com/publications/abstract.php?pid=919

Wietze, L. and Bakker, K. (2005) 'Economic regulation of the water supply industry in the UK: a game theoretic consideration of the implications for responding to drought risk', *International Journal of Water*, 3(1).

Willner, J. (2001) 'Ownership, efficiency, and political interference', *European Journal of Political* Economy, 17: 723–48.

Willner, J. and Parker, D. (2007) 'The performance of public and private enterprise under conditions of active and passive ownership and competition and monopoly', *Journal of Economics* 90(3), April.

4
The Electricity Industry Reform Paradigm in the European Union: Testing the Impact on Consumers

Carlo V. Fiorio, Massimo Florio, and Raffaele Doronzo

University of Milan

Abstract

The standard electricity industry reform paradigm in several EU countries since the 1990s includes privatization, unbundling, liberalization. While the implementation and design of reforms widely differ across the EU, the European Commission insists on a rather unified approach, aiming at the full opening of the internal market. Privatization neither is a necessary pre-requisite of liberalization, nor is it mentioned in EU electricity market directives. Many economists, however, believe that public ownership can be an impediment to other reforms, and that it leads to production inefficiency. To test the latter question and the reform paradigm in general, as captured by a set of regulatory indicators, we consider electricity prices and survey data on consumer satisfaction in the EU-15. Our empirical findings reject the prediction that privatization leads to lower prices, or to increased consumer satisfaction. Country-specific features tend to have a high explanatory power. The progress toward the reform paradigm is not systematically associated with lower prices and higher consumers' satisfaction. We discuss possible interpretations of our findings, suggest possible explanations, and some policy implications.

Keywords: Electricity industry, privatization, liberalization, unbundling

Journal of Economic Literature classification: L94, L33, L43

Acknowledgments: The research has been supported by the PRIN 2005 Programme (prot. 2005137835): 'Fiscal and Regulatory Challenges of the

European Integration: an Agenda 2007–2013'. The authors are grateful to Giuseppe Bognetti, to participants in the Understanding Privatization Policy meeting in Prague, February 2007, the European Economy Workshop in Milan, June 2007, the International Conference Developments in Economic Theory and Policy in Bilbao, July 2007, the XIX Public Economics Conference in Pavia, September 2007, for helpful comments, and to Giancarlo Manzi for competent research assistance.

1. Introduction

This chapter offers an empirical analysis of the effects of privatization and liberalization on prices in the electricity industry. It begins by locating the debates on the effects of privatization in the context of the electricity industry. It then reviews some features of the electricity industry and its reform in selected EU countries. Then we offer an empirical analysis of country panel data on prices. The subsequent section considers evidence based on survey data relating to consumer satisfaction with prices and quality and proposes simple empirical models. A subsequent section sums-up and discusses our findings. The chapter is concluded by a reconsideration of the paradigm and some policy implications.

2. Electricity industry and the privatization debates

The electricity industry can be described as including four different activities: generation, transmission (the high voltage network), distribution (the middle and low voltage network), retail (supply to final consumers). These four activities or industry segments have different technological and economic characteristics. Generation is often considered potentially competitive, because economies of scale in most types of production process are said to be not large. Transmission and distribution are natural monopolies, at the national and regional level, because of the high network fixed sunk costs. Eventually, retail supply is often seen as potentially competitive, because trading and marketing activities do not imply high fixed costs.

Traditionally, all or some of these activities were vertically integrated in many European countries, with state or municipally owned enterprises playing an important role. The market was highly regulated, with very limited opportunities for users to switch to alternative suppliers. There was no third party access to the transmission grid. This integrated pattern was the deliberate result of policy reforms that consolidated the mostly private and fragmented European electricity industry in its earlier stages, under the governments' view that for economic, political and social reasons the previous pattern, mainly based on regional private monopolies or collusive oligopoly, was either inefficient or undesirable (see Millward, 2005 for a detailed history of nationalization and consolidation in Europe). Despite some policy convergence in the interwar years, and further reforms in the 1950s, in the following half century there were however persistent and significant variations in industry patterns across countries in Europe, in terms of technologies, ownership, governance, per capita-consumption, spatial and vertical integration, market structure, and prices.

Following privatization and liberalization in the UK in the 1990s, and the EU directives in the last ten years, a new paradigm has emerged, or 'a measure of consensus over some generic measures for achieving a well functioning market-oriented industry' (Jamasb and Pollitt, 2005, p.2). For the first time in the history of the electricity industry in Europe a unique cross-country policy reform pattern has been advocated by international organizations, notably the European Commission, the OECD, and the World Bank in the transition economies.

The new paradigm is usually simplified as suggesting three parallel reforms: privatization (sale of existing publicly owned firms and licensing of private entrants), unbundling (associated with incentive regulation of the networks, third-party access, establishing an independent regulator), and liberalization (i.e. allowing entry and competition in generation and retail).

The World Bank (2007) suggests a quite longer list of reform items:

a) De-monopolization and regulation: unbundling vertically integrated monopolies to foster competition in generation and supply; privatize and shifting the role of the state from owner to regulator; promoting entry by foreign investors; establishing transparent energy markets; building regulatory capacity
b) Prices and fiscal policy: promote fully cost-reflective prices; elimination of production subsidies; taxation based on externality correction; enforce metering and collection of bills; closing uneconomic plants
c) Foreign trade: open markets to imports; eliminating taxes on fuels and electricity; strengthening regional trading arrangements; expanding transboundary energy connections
d) Investment policy: rely on energy companies to sustain investment, not on the public sector budget; support energy efficiency; increase flows of foreign capital with appropriate measures

e) Social protection: safety nets for the redundant staff; social service functions to be transferred to local governments, not to companies; support to the poor through lifeline tariffs or means-tested subsidies (abolish cross-subsidies)
f) Environmental protection: supporting environmental assessment; introducing emission norms; mainstreaming new environmentally friendly technologies.

Looking at this comprehensive package of policy reforms it seems that many items are only loosely interrelated and can be implemented under a variety of industry structures and government interventions, thus the degrees of freedom in the reform design are higher than sometimes is suggested. For example, opening market to imports may be consistent with different ownership regimes: in principle, there is nothing that prevents the import of electricity from abroad when the generation or the transmission network is publicly owned. Vested interests against international trade can be strong either under public or private ownership. Unbundling can be legal, accounting, or ownership separation, with quite different implications. Promoting entry of foreign investors is compatible with weak promotion of transboundary physical networks; in fact it would amount to offering rents because of distorted domestic prices to a coalition of investors larger than the national one. Vertical integration can be combined with liberalization under effective third-party access. Liberalization of generation can be combined with constraints to retail competition. It is thus difficult to see the logical necessity of linking together all the items of the reform package. The case for implementing the comprehensive paradigm rests on a mixture of beliefs and evidence, with only limited cross-country empirical research based on standard econometric approaches.

Without empirical testing, however, some of the tenets of the reform paradigm are questionable, or at least depend upon a large number of pre-conditions. For example, while sometimes the new paradigm has been justified by dramatic technological changes that were assumed to reshape the industry, e.g. new generation process using gas as fuel and the loss of economies of scale, this argument seems to be simplistic. The mix of energy sources in Europe is too diverse to confirm this explanation. Nothing of importance has changed in most generation processes, from nuclear to hydro, in the last 20 years in most countries, to justify an overwhelming technological explanation of the paradigm shift.

It seems more reasonable to look at the new paradigm as a set of policy reforms based on increased confidence in market forces and private ownership, against the decline in confidence in planning and public ownership, for a number of reasons that we do not discuss here. Moreover, the ambition by the EC to create an 'internal' market for services, may contribute to the explanation of the policy reversal, more than any compelling technological shift.

In this perspective, the driving idea behind privatization of electricity companies is that public ownership is intrinsically less efficient than private ownership, because of an incentive argument. In turn, the rationale for unbundling is to separate the potentially competitive stages from those with natural monopoly characteristics, which may need some public regulation. Eventually, liberalization would bring market forces into the industry, and competition would deliver production and allocative efficiency, hence lower prices, or lower mark-up over costs, for users.

As mentioned, while the consensus on the new paradigm is high in the EU, the empirical evidence that supports it is more limited. After more than ten years of experience with its implementation, it seems appropriate to move from speculation on its merits to testing its impact on empirical grounds. Particularly, it would be interesting to check the differential impact of privatization against other reforms, because admittedly their association is far from granted, see Jamasb and Pollitt, 2005. In fact, while the EU directives are mute on this point, OECD economists tend to link together ownership and market reforms (see Conway and Nicoletti, 2006; or Lampietti *et al.*, 2007).

Empirical research on the impact of electricity reform at national level is available in some EU countries, but a major problem in this area is often the lack of appropriate analysis of counterfactuals, i.e. comparing scenarios with and without reforms, or with a different mix of reforms (Newbery and Pollitt, 1997). The data needed to perform this analysis are often not easily available. The large variations in progress in the implementation of the new paradigm across the EU countries, while a matter of disappointment for the EC, offers however a proxy for such counterfactuals. By observing different industry characteristics in different countries we can try to infer the differential impact of reforms.

Thus, our empirical approach is to take advantage of the diversity in European electricity reform patterns and to control for a number of potential explanatory variables to predict two simple performance indicators: prices of electricity to consumers, and satisfaction of consumers with prices they pay and quality of service provided. As for prices, we use standard statistical databases, such as those provided by Eurostat and the International Energy Agency (IEA). For consumer satisfaction we use three waves of the Eurobarometer survey (2000–2002–2004), a well-known social attitudes study performed on behalf of the European Commission. To describe the national reform patterns and trends we use REGREF, a database developed by the OECD.

This combination of objective and subjective evidence allows us to double check our findings. Panel analysis of price trends, using regulatory and technological variables, plus country macroeconomic and other structural features, offers objective evidence on the observed impact of reforms at an aggregate level. Our findings can be compared with the results of the main study commissioned by the EC, DG Internal Market, see Copenhagen Economics (2005). Micro-data on consumers perceptions capture additional information, not covered by aggregate data, and not considered in detailed in the Copenhagen Economics Study.

The reason for cross-checking objective and subjective evidence is also that, as revealed by Lampietti *et al.* (2007) in the context of the transition economies, aggregate data in same cases may conceal important issues, such as tariff rebalancing, social affordability, quality of services (e.g. interruption or irregularity of voltage), non-payment, shifts to other sources. While we cannot go in depth as they do in their country studies, our analysis by using a large survey dataset (Eurobarometer) is, as far as we know, the first attempt to see how consumers in the EU-15 perceive the price they pay and the quality of service, conditional to the reform variables and a number of individual and country features.

Our main findings are the following: first, panel estimation of prices tend to reject the prediction that privatization per se leads to lower electricity prices, after controlling for other reforms, and other industry and country-specific variables; second, customer satisfaction for prices is correlated to observed prices, confirming that perceptions by consumers are broadly consistent with the objective evidence; third, customer satisfaction about prices and quality of services is higher with public ownership than under private ownership.

Moreover, in general, country-specific fixed effects tend to dominate the explanations compared with the regulatory variables. We conclude that the evidence, at least at this stage of reforms, does not (yet?) support the view that a unique paradigm of privatization-unbundling-liberalization is clearly more beneficial to electricity consumers in Europe than others.

3. European industry reforms in the EU-15

3.1. The EU directives

The aim of the first EU electricity directive of 1996[1] was to gradually introduce competition in order to create a European market for electricity. Some principles were established: the unbundling of different production stages, the introduction of transparent rules for licensing new generation plants, the gradual extension to final customers of the right to buy electricity directly from the producers and the right of access to the network by new entrants. From July 2004 a new directive repealed that of 1996.[2] Aim of this new directive is to speed up the integration process and the development of competition and therefore it is more detailed than the old one on several issues, the most relevant of which are clearer rules about the unbundling of production stages (especially the separation of the grid from the supply); the compulsory creation of a national energy regulator; and the immediate opening of the market to all customers. In particular, the directive requires market opening to all non-household customers by 1 July 2004 and to all customers by 1 July 2007.

This new Electricity Directive was due to be transposed in national legislation by Member States by July 2004. These requirements aim to yield a strongly market-led system. However, several Member States have been slow in implementing these measures. At the end of 2005 Greece, Luxembourg, Portugal and Spain still did not fully notify the Commission of the legal measures taken for the purposes of transposition. In November 2006 the European Commission published the report on the functioning of the internal market in electricity and gas. This report confirmed that cross-border competition was not yet sufficiently developed to provide customers with a real alternative from the nationally established suppliers. Key indicators in this respect were the absence of price convergence across the EU and the low level of cross-border trade. This situation was a direct consequence of the failure of Member States to implement the second electricity directives on time or with sufficient determination. Furthermore, inadequate use of existing infrastructure and insufficient interconnection between many Member States prevents real competition from developing, despite the political commitment of the European Council made in 2002 to achieve an import capacity of at least 10 per cent of internal consumption. Out of the 15 Member States before the 2004 and 2007 enlargements of the EU, in 11 there are companies that have strong or dominant market positions. In some cases, far from reducing their market power, these companies now appear to have more market power than they did before the first directive was passed.

To illustrate the diversity of the industry in the EU we briefly sum up here some features of the industry in selected countries. This is not intended to be a comprehensive account, rather just a way to stress the persistence of structural diversity in industry patterns across EU countries.

3.2. Electricity supply sources

A shortcut way to describe the electricity generation side of the electricity industry in any country is to look at the sources of energy. Geography and national strategies have shaped in the EU a very uneven landscape. In this subsection, and in the following ones in this section of the chapter, we focus on five major countries in the EU-15: UK, France, Germany, Italy and Spain, and in three Nordic smaller countries: Sweden, Denmark, Finland.[3]

The productive mix in the UK, around 2002–04 comprises roughly 80 per cent fossil fuels (gas and coal in equal shares, very modest role of oil) and the remaining part is almost entirely provided by nuclear power (hydro, wind, other renewables play a marginal role). This pattern can be compared with France: in 2004 here almost 78 per cent of the electricity produced was generated by nuclear plants, only 10 per cent from fossil fuels, and 11.5 per cent from hydroelectric and renewable sources. This situation leads to much lower direct (short run) production costs than in the other EU-15 countries.

The Spanish productive mix is more balanced. There is a prevalence of fossil fuels (59 per cent, more than half of it being coal) followed by nuclear (22 per cent), hydroelectric (12 per cent) and other renewables (6 per cent). The pattern in Italy is very different, where the current productive mix is the result of a political choice (following a referendum) to de-commissioning nuclear power stations. In 2004 80 per cent of the electricity was derived from fossil fuels, with oil and gas each around 30 per cent, and 16.5 per cent is of hydroelectric origin, while only 2.6 per cent is from other renewable sources. Germany has a productive mix more similar to Spain, which includes 63 per cent fossil fuels (but mainly coal), 27 per cent nuclear and the remaining 10 per cent produced by hydroelectric and other renewable sources.

Turning to the Nordic countries, differences in their energy sources for electricity are also impressive. In 2004 the energy in Sweden is produced mainly by nuclear (50 per cent) and hydro (39 per cent) with only 10 per cent coming from fossil fuels. In Finland, the largest source of electricity is the fossil fuels (55 per cent) but a relevant part is derived also from nuclear (26 per cent), with the remaining from hydro and other renewable sources. As regards Denmark, the productive mix is dominated by fossil fuels (83 per cent, mainly coal) with the remaining electricity entirely derived from renewable sources different from hydroelectric.

3.3. External trade

International trade of energy is marginal in most countries, but not in all. The UK is marginally a net importer of electricity. In 2004 its imports satisfied around 2 per cent of demand. Spain is a marginal net electricity exporter in 2004 (+1 per cent). France, in contrast, is the most important European net exporter of electricity. Roughly 15 per cent of the electricity produced in the country is sold to Italy, Germany, UK, Spain and Belgium. Germany is also a net exporter of electricity, while Italy is a substantial net importer of electricity: roughly 13 per cent of demand is satisfied by imports from Switzerland, France, Slovenia and Austria.

Looking at the Nordic countries, Sweden imports only a small fraction of its internal demand of electricity (1.4 per cent in 2004), but Finland imports almost 5 per cent of its demand in 2004, and Denmark exports a substantial fraction of its production (almost 8 per cent in 2004).

It is interesting to record here that there are other EU countries with substantial international trade of electricity: Luxembourg, Netherlands, Belgium import more than 20 per cent of their consumption. Austria and, outside the EU, Switzerland have established themselves as active electricity traders.

3.4. Reform trends

The UK was the first European country to implement a comprehensive reform of the electricity sector at the end of the 1980s. The liberalization process started in 1989 with the Electricity Act. It was completed in 1999. The reform contemplated the liberalization and simultaneous privatization of the two state companies (National Power and Power Gen), and of a network of RECs (regional electricity companies). The privatization process ended in 1995, in England and Wales, but Scotland and Northern Ireland followed a different path. As regards the market opening at the beginning the right to choose the electricity supplier was granted only for the major users, but the threshold level was progressively reduced, reaching a complete liberalization in 1998. The Electricity Act established also the introduction of a wholesale market and the complete separation of the transmission activity from generation. In the network services (transmission and distribution) the price was regulated by the price cap, under the control of OFFER, a regulation agency, now called OFGEM.[4]

Again in contrast to the UK, France was one of the latecomers in implementing the EU directives. The experience of liberalization began only in

2000 when the government approved the law n. 108 which modified the pre-existing structure created by the Law of 1946. The reform predicted a regime of regulated third-party access, the introduction of a wholesale market and a progressive opening of the sector with the possibility for the consumer to choose the retailer. The threshold level was established in 100 GWh in 2003 and was gradually reduced in the subsequent years and finally removed in 2007.[5]

The reform of the energy sector began in Spain with the privatization of two public electric utilities. From 1988 to 1999 the privatization of *Red Electrica* and *Endesa* was completed. In 1994 the Law 40/1994 known by the acronym LOSEN was enacted with the aim of liberalizing the sector. This law mandated the legal unbundling of the transmission network and created an independent joint public–private TSO[6] called *Red Electrica Espanola* (REE). REE offers regulated TPA[7] to both the transmission network and the distribution network. In so far as generation LOSEN was quite permissive in terms of the maximum market share that any given Spanish generating company could control. Therefore the subsequent years were characterized by an acceleration of the trend towards consolidation: from over 35 independent regional generation companies in 1990 only five were left by 2002. The LOSEN was surpassed by Law 54/1997 which accelerated the process of liberalization. Full market opening, including for domestic customers, and regulated third-party access, have been implemented since January 2003. An energy regulator was also created: the *Comisión Nacional de Energía* (CNE). However, important elements of the Directives have not been transposed, and the adoption of the implementing legislation has constantly been delayed for the last two years. Spain is therefore the only other Member State with Luxembourg, subject to general infringement procedure for non communication of transposition measures for both gas and electricity Directives, that are now before the European Court of Justice.[8]

In Italy, the transposition of the first EU directive was realized by the implementation of the Bersani decree in 1999. The main changes implemented were the introduction of competition in generation and supply to end consumers; the vertical separation of the production stages; and the creation of a wholesale market. The reform of 1999 did not provide for an immediate full market opening: from 2000 those with a consumption of over 20 GWh have been able to join the free market. Since then the threshold has been progressively reduced to achieve total liberalization in 2007. The Bersani decree also imposed the vertical disintegration of ENEL spa with the creation of TERNA. This new society controls almost the entire transmission system and is wholly owned by

ENEL. The management of the grid was devolved to GRTN (*Gestore della Rete di Trasmissione Nazionale*), a publicly owned society. The GRTN guarantees open access to the network, according to the conditions and tariffs determined by the regulator. A decree of May 2004 approved the reunification of the transmission network's management and ownership. This was done to eliminate co-ordination problems between GRTN and TERNA. In June 2004, 50 per cent of TERNA's capital was sold to the public for €1.7 billion. To ensure the independence of the new enterprise TERNA Rete Elettrica nazionale spa, from 1 July 2007 non-transmission energy companies are prohibited from owning more than 20 per cent of the shares of the new combined company (voting rights must not be higher than 5 per cent). The same decree entrusts TERNA with all of GRTN's responsibilities.[9]

Germany began to implement the EU directives in 1998 with the Energy Act and completed the reform in 2005 with the Energy Industry Act. The reform introduced full market opening: each consumer had the right to choose his own supplier without any limitations or volume threshold. The supplier could also be a producer, a regional distribution company or a municipalized company. An electricity exchange was also set up, but only in the August of 2005 a regulatory body, the Federal Net Agency, was installed with the mandate to control the prices of the industry.[10]

In 1996 there was a radical reform in Sweden following the experience of Norway.[11] After the enacting of the Law for Electricity Supply total production liberalization and a regime of authorization were implemented. The transmission system remained in public hands, in a monopoly regime managed by a non-profit state company, and distribution was delegated to 280 municipalized firms. The retail market was fully opened and since 1996 consumers can choose their own supplier without any threshold level. The separation between transmission and distribution operators has been effective since 1992, but only in compliance with the second EU electricity directive the overall degree of vertical integration of the industry became completely unbundled.[12]

Finland decided to reform the sector in 1995 with the Electricity Market Act. From then on the market was progressively liberalized and since 1997 the consumer can freely choose her/his own supplier. The reform also establishes the division between the transmission and generation firms although the distribution companies are not yet fully unbundled.[13]

The reforming experience in Denmark is very similar to that of the other Nordic countries and began in 1996 with the Danish Supply Act and was completed in 2001 when the law established the complete rights

Table 4.1 The EU-15 electricity liberalization laws

Country	Law
Austria	Law of electricity supply 1998 – Energy Regulator Act 2000
Belgium	Law for the organization of the electricity market 1999 – Federal law of 06/05
Denmark	Danish supply Act 1996 – Act No. 375 of 1999 – Act No. 138 of 2004
Finland	Electricity market Act 1995 – Amendment to the Electricity Market Act 2004
France	Law n. 108 of 2000
Germany	Act on the supply of Electricity and Gas 1998 – Energy Industry Act 2005
Greece	Electricity Law 1999
Ireland	Electricity Regulation Act 1999 – Utilities Act 2000
Italy	Bersani decree 1999 – Marzano Law 2004
Luxembourg	Law on the organization of electricity market 2000
Netherland	Electricity Act 1998
Portugal	Decree Laws 182–85 of 1995
Spain	Law 407-1994 Electricity Act (LOSEN) – Law 54/1997 (Electricity Power Act)
Sweden	Law for Electricity supply 1995
UK	Electricity Act 1989

Source: Commission of the European Community (2006) and Al-Sunaidy and Green (2006).

for any type of consumer to choose the supplier. The electricity sector is characterized by the presence of public local companies and despite some privatization the overall ownership structure still remains mostly public. The separation between generation and transmission activities was established in 1996, but the total unbundling of the industry, also regarding distribution, was completed in 2001.[14] Tables 4.1 and 4.2 show some of the acts implementing the EU directives and the time table.

3.5. Market structure

There is no evidence that market structures are converging to a unique pattern in the EU, even if there are some common trends driven by the policy initiative of the European Commission.

After the process of liberalization the number of companies producing electricity in the UK increased considerably from six to 47 in two years. More recently however there have been processes of re-integration, mergers and acquisitions that have increased the concentration of the market. Generation consists of approximately 20 private entities, with only seven

Table 4.2 Timetable of the liberalization process in EU-15

	89	90	91	92	93	94	95	96	97	98	99	00	01	02	03	04	05
Austria									x		x						
Belgium										x							x
Denmark							x			x					x		
Finland																	
France											x						
Germany								x									x
Greece										x							
Ireland										x	x						
Italy										x					x		
Luxembourg										x							
Netherland									x								
Portugal						x											
Spain					x		x										
Sweden						x											
UK	x																

Source: Commission of the European Community (2006) and Al-Sunaidy and Green (2006).

having a market share above 5 per cent. In particular the market is dominated by Innogy (21 per cent), British Energy (18 per cent) and PowerGen (17 per cent). The retail market is characterized by the presence of seven big suppliers with a market share above 5 per cent, while the number of distributors is higher (15 distribution companies in 2004). The ownership of the transmission grid is divided between four different private companies. In the UK there are many 'broker' type services that act to help customers choose the best contract structure offered by the main suppliers. In summary, for electricity there would appear to be a sufficient range of companies to suggest that the market is both competitive and open to new entrants. The main development in recent years is the greater degree of integration between the major producers and supply companies.[15]

In France the structure of the industry is very different from the UK. The French national government has a strong traditional relationship with its energy industry. The primary electric utility *Electricitè de France* (EDF), is a vertical integrated public company formed through a process of nationalization of the electric energy industry by the law of 1946. As regards unbundling the government decided to introduce only a form of accounting separation between transmission and generation, essentially maintaining the vertical integration of the electric sector. Another characteristic of the process was the reluctance to privatize EDF: the state in 2006 controls 70 per cent of his capital. Although new private generators have been given access to the market, EDF still owns and controls the majority of generation as well as the transmission and distribution networks. EDF is responsible for approximately 87 per cent of the generation capacity in 2005, while the *Compagnie nationale du Rhone* (CNR), also a publicly owned company, Electrabel-Suez and SNET (group Espanol-Endesa) control almost entirely the remaining share. In 2001 CNR created *Energie du Rhone* as a joint venture with Electrabel, the former Belgian national utility, to distribute and retail the electricity CNR produces in France. RTE, the subsidiary company of EDF created in 2005, is the unique administrator of the transmission system.[16]

In Spain, the most relevant outcome of the electricity policy so far has been the emergence of some big firms that consolidated their generation assets. The number of firms with a market share above 5 per cent in this stage is particularly low: the market is overwhelmingly controlled by the three largest companies Endesa, Iberdrola and Union Fenosa, which together account for 83 per cent of the generation installed capacity. The competition in the retail market contrasts greatly, with almost 300 firms operating, only 3 having a market share above 5 per cent in 2004. In July 2006 the Iberian Electricity Market, MIBEL, began operating. It aims at creating an integrated electricity wholesale market with Portugal, notably by creating a single market operator for the wholesale Iberian pool market but, despite the advent of MIBEL, it is too soon to assume that a unified Iberian market exists.[17]

After a long history of private oligopoly, Italy changed to a vertically integrated public monopoly in 1963. ENEL, the listed public sector company, became a public limited company in 1992. After the privatization process begun at the end of the 1990s, the government now controls 21 per cent of the company. In generation the decree allowed ENEL to retain a maximum market share of 50 per cent after 2003. As a consequence, three of its generation companies were sold on the market: the sale of Elettrogen, Eurogen, and Interpower were completed by the end of 2003. In terms of electricity generated five operators have a market share over 5 per cent in 2004: ENEL Produzione (39 per cent), Edison group (11 per cent), ENI Group (9 per cent), Endesa Italia (8.2 per cent) and Edipower (8 per cent). The distribution network is again virtually controlled by ENEL, with the exception of a few cities where the local municipalities own the distribution companies.[18]

In Germany, prior to the recent reform, the law of 1935 had led to the existence of a de facto regime of private regional monopoly with nine vertically integrated supra-regional companies, which in 2000 still controlled 80 per cent of production, 40 per cent of distribution and all transmission. Below this oligopolistic level that has been static for decades, a regional level operates which is formed of about 80 firms whose principal activity consists of acquiring energy from the large distributors and reselling it at a local level. About 800 entities operate at a local level, most of them publicly owned. With respect to generation, 90 per cent of the total electricity in Germany is produced by only four large groups: ENBW, Vattenfall Europe, RWE and E.ON. In terms of electricity supply each of the four large companies holds a market share of much more than 5 per cent and the market share of another six companies ranges from 0.9 per cent to 2.1 per cent. Together these ten companies supply more than 60 per cent of total German electricity consumption, while private generators and municipal utilities mainly make up the remaining. As for the separation between the transmission and the generation process, despite the two stages being officially divided since 2002, the European Commission argues that there is still insufficient unbundling because the interest structures of the transmission system operator seems to be influenced by the supply interest of incumbent companies. The transmission assets are split between four companies while the number of distributors is particularly high. The ownership structure of the industry continues to be mixed with the contemporaneous presence of public and private operators. To sum up, the German electricity market is still characterized by a high degree of vertical and horizontal integration dominated by very few large companies. This structure in combination with congestion at interconnectors as well as some specific problems related to network access is thought to prevent effective competition from developing and to increase barriers for new entrants and independent investments in new power plants.[19]

The three Nordic countries, despite being well integrated in one regional market do not show structural similarities. A few companies dominate the production of electricity in Sweden. In 2005 the former monopolist Vattenfall Fortum, still 100 per cent government owned, and Eon Sweden accounted for 88 per cent of the country's electricity production. Supply companies and big end customers are able to buy electricity from the wholesale market. Customer switching is very common in the Swedish retail electricity market. In total 54 per cent of electricity customers have either renegotiated their contracts or switched supplier since the market reform in 1996. Generally speaking competition on end customers is considered to work.[20]

In Finland the generation market is dominated by two big players, namely Fortum (majority state-owned) and PVO. There are moreover several smaller producers on the market, including municipal companies, often publicly owned. The overall degree of competition in the country after the reform process is satisfactory: there are five companies with a market share above 5 per cent in the generation stage, while the structure of the retail market is much more concentrated with three firms having a market share above 5 per cent. Customer switching is reasonably common in Finland. Among small and medium-sized industrial and business customers, it is estimated that 82 per cent of the volume and 50 per cent of the number of customers of electricity bought by customers is based on contracts that the customers have either renegotiated with their local supplier or made with a new supplier.[21]

After the reorganization of the energy sector, the newly restructured state-owned electricity and gas company DONG Energy now dominates the generation market in Denmark. This former gas company has acquired most of the electricity generation sector. In the reorganization a few large power plants were sold to Vattenfall, the Swedish state-owned electricity company. The generation market thus remains oligopolistic with two main generating companies and numerous smaller generation companies with distributed generation assets. There is some competition from Nordic and German generators through the substantial interconnection capacity. In the retail segment of the industry there is a great number of suppliers (more than 100 in 2003) with five having a market share greater than 5 per cent. Customer switching is relatively common in Denmark for industrial customers: more than 50 per cent of the customers are estimated to have either renegotiated their contracts or switched suppliers. In contrast, the retail market for small customers remains de facto closed due to the regulated default supplier tariffs.[22]

All the Nordic countries are part of a joint wholesale electricity market. The Nord Pool market was created in 1996 by Norway and Sweden and afterwards was completed with the entry of Finland (1998) and Denmark (1999). So the profile of the electricity industry of all these countries has to be considered jointly. If we look at the entire Nord Pool we can discern a competitive context: the generating companies with a market share over 5 per cent are 10, while the aggregate share of the three largest generators is under 40 per cent, and this seems to be compatible with mainly public ownership of generation and of the networks.

In general in the EU concentration in the industry remains high, with the largest three generation firms or the three largest retailers controlling more than 60 per cent of the market in the large majority of countries, whatever the extent of privatization, with two polar exceptions: the UK and the Nordic countries, the former with no public ownership left, the latter with mostly public sector firms.

This brief overview of the electricity industry in five major EU countries, and in three smaller Nordic countries, shows a striking variety of patterns. As mentioned, some common trends can be discerned but it is not evident that a unique European pattern is emerging. This, in a sense, is good news for empirical analysis, because it allows us to test whether structural diversity is associated with different performances. We try in the next sections to use available comparative data to test the reform paradigm in the EU-15. We consider this as exploratory research because we are fully aware of the difficulty of capturing the structural and legal diversity mentioned in this section. There are however very few attempts in the literature to evaluate the electricity reforms across the EU, and it seems useful to contribute to the policy reform debate on empirical grounds.

4. Data used for price analysis

The first source of data for this chapter is the IEA 2006 Electricity Information data set, which collects data about electricity variables of OECD countries (IEA, 2006). The time series of gross electricity prices and electricity tax for households, which will be extensively used here, start in 1978 and for most of the EU-15 countries it ends in year 2005. Figures 4.1 and 4.2 plot the log prices of electricity for households since 1978 for the EU-15 countries, measured in purchasing power parity (PPP) per Kwatt/hour. They show that log prices have increased with similar trends throughout the period showing some convergence of rate of growth at the end of the period with respect in particular to mid-1980s for all countries considered. The data set used comprises 408 observations on household prices across the EU-15 but it presents several missing observations, especially in the cost variables (less than 200 observations are available for gas cost variable).

We also extensively use REGREF, an OECD regulatory database (Conway and Nicoletti, 2006), which collects some indicators of privatization, vertical disintegration, market liberalization of several services of general interest across some OECD countries. Focusing on the electricity market we use in particular the variable 'public ownership', which measures the public ownership of each SGI and is coded from 0 (private ownership) to 6 (public ownership), the variable 'vertical integration', which is an indicator of vertical separation in different industries and is coded from

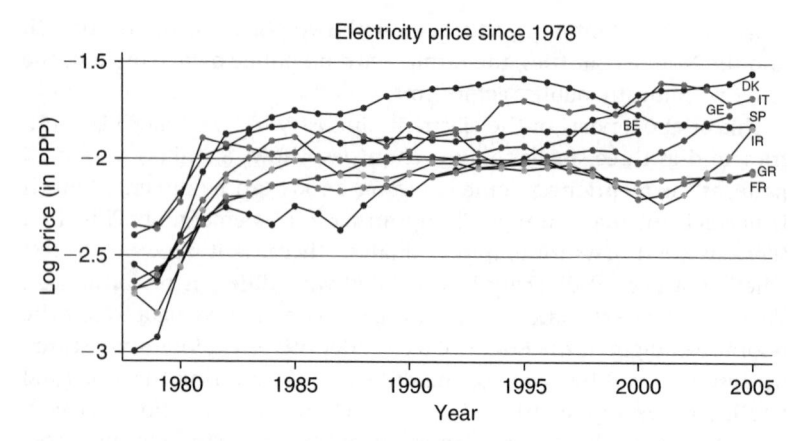

Figure 4.1 Electricity (log) price dynamics for households, in PPP per *KW/Hour*
Source: Our calculations on IEA (2006).
Notes: BE = Belgium, DK = Denmark, GE = Germany, GR = Greece, IT = Italy, SP = Spain, FR = France IR = Ireland. National currency is euro, except Denmark.

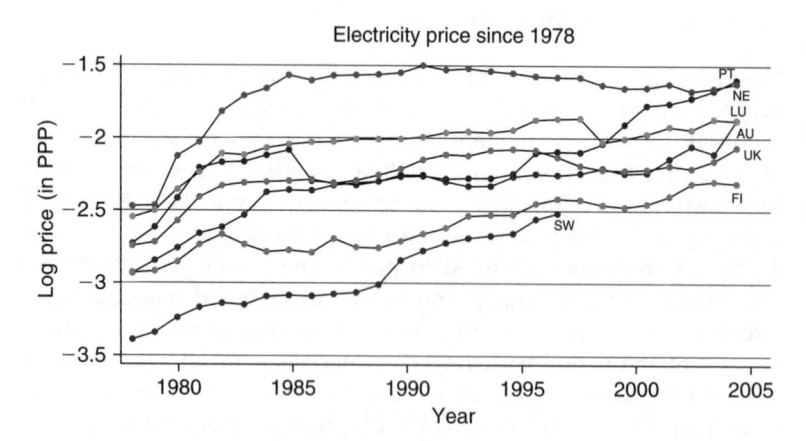

Figure 4.2 Electricity (log) price dynamics for households, in PPP per KW/Hour
Source: Our calculations on IEA (2006).
Note: LU = Luxembourg, NE = Netherlands, PT = Portugal, UK = United Kingdom, FI = Finland SW = Sweden, AU = Austria. National currency is euro, except Sweden and UK.

0 (ownership separation) to 6 (integration), and the variable 'entry regulation', which is a weighted average of legal conditions of entry in a market and is coded from 0 (free entry) to 6 (franchised to one firm). Although in some cases these variables only take discrete values, they are allowed to take any value in the 0–6 range (Table 4.3) and at present

Table 4.3 Some regulatory indices about the electricity industry

	Public ownership			Entry regulation			Vertical integration		
	1999	2001	2003	1999	2001	2003	1999	2001	2003
Belgium	1.50	1.50	1.50	5.00	2.30	2.30	6.00	1.50	0.00
Denmark	3.00	3.00	3.00	0.30	0.30	0.00	6.00	1.50	0.00
Germany	3.00	3.00	3.00	3.00	1.00	1.00	3.00	3.00	1.50
Greece	6.00	6.00	6.00	6.00	2.30	2.30	4.50	4.50	1.50
Italy	6.00	3.00	3.00	4.00	2.30	0.30	4.50	1.50	0.00
Spain	1.50	1.50	1.50	0.30	0.30	0.00	1.50	1.50	0.00
France	6.00	6.00	6.00	4.30	2.30	0.30	6.00	4.50	4.50
Ireland	6.00	6.00	6.00	4.80	2.30	2.30	4.50	1.50	1.50
Luxembourg			0.00						4.50
Netherlands	6.00	0.00	0.00	0.30	0.30	0.30	1.50	1.50	1.50
Portugal	3.00	3.00	3.00	4.30	2.30	2.30	3.00	1.50	1.50
Great Britain	0.00	0.00	0.00	0.00	0.00	0.00	0.00	0.00	0.00
Finland	3.00	3.00	3.00	0.00	0.00	0.00	1.50	1.50	1.50
Sweden	3.00	3.00	3.00	0.00	0.00	0.00	1.50	1.50	0.00
Austria	3.00	3.00	3.00	4.00	2.00	0.00	4.50	1.50	1.50

Source: our calculations on REGREF (Conway and Nicoletti, 2006).

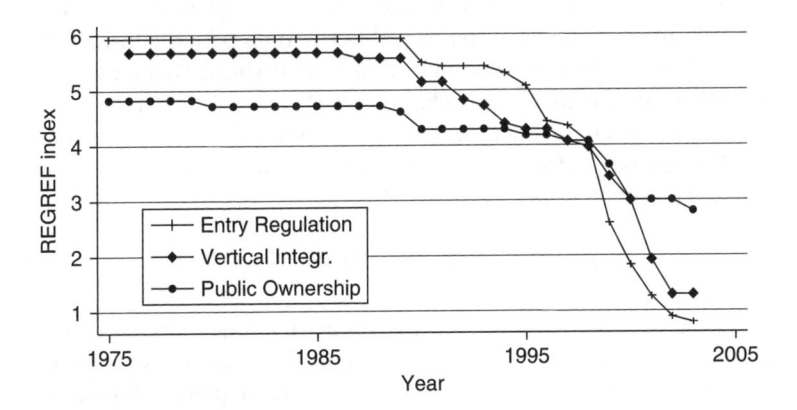

Figure 4.3 Trends of mean of regulatory indices across EU-15
Source: our calculations on REGREF (Conway and Nicoletti, 2006).

the time series starts in 1975 and ends in 2003. The common trend across the EU-15 countries since 1985 has been towards a marked reduction of public ownership, a less integrated industry structure and less regulated access to the market (Figure 4.3).

5. Explaining electricity price dynamics

In this section the main hypothesis we test is whether log prices are affected by regulatory variables, such as vertical integration, public ownership and entry regulation. Let p_{it} be the log of household (net of taxes) electricity prices for country i at time t, R_{it} the vector of regulatory variables for country i at time t, which includes vertical integration, public ownership and entry regulation, t the deterministic trend and X a set of controls including production costs, residential consumption, efficiency losses, GDP, national population, imports and type of energy source. We then model log prices as:

$$p_{i,t} = c + R'_{i,t}\beta + X'_{i,t}\phi + \alpha_i + t\gamma + \varepsilon_{i,t} \qquad (1)$$

where $c, \alpha, \beta, \gamma, \phi$ are parameters and ε is the error term. Model (1) presents a couple of main limitations. First, it overlooks the possible presence of common trends among dependent and independent variables. Second, it is not a dynamic model. The first problem relates to the fact that since 1978, which is the starting date of our time series, electricity (log) prices presented an upward trend, although different across countries. Analogously, other variables, such as regulatory indices and cost variables also showed trends across the time period considered. A Dickey–Fuller test for the presence of unit roots shows that all variables considered are integrated of order one, with the only exception of the population variable, which presents a deterministic trend.[23] Hence, to avoid the problem of spurious correlations, we estimated the model with stationary variables only:

$$D \cdot p_{i,t} = c + (D \cdot R_{i,t})'\beta + (D \cdot X_{i,t})'\phi + \alpha_i + \delta_{i,t} \qquad (2)$$

where the symbol $D\cdot$ before a variable shows that it has been differenced, hence made stationary. Estimation of model (2) using fixed-effects is presented in Table 4.4, column (A). Fixed effects are preferred to random effects as p_{it} is conditional on α_i, where α_i can be estimated and can reasonably be interpreted as 'one of a kind' and cannot be viewed as a random draw from some population, since it represents countries.

Model (2) was then augmented to include other variables, such as energy sources, cost and macroeconomic variables. Table 4.5 shows that the impact of regulation (as measured by the REGREF indicators) on the change of log electricity prices is rather small and not statistically significant. Although someone may argue that the length of the time series

Table 4.4 Fixed-effect panel estimation
Dependent variable is log net (of tax) price of electricity for households

	Panel regression with stationary variables, fixed effects		
	(A)	(B)	(C)
D.Vertical Integr.	−0.001	0.001	0.007
	[0.011]	[0.006]	[0.013]
D.Public Ownership	0.011	−0.005	0.017
	[0.014]	[0.008]	[0.018]
D.Entry Regulation	0.006	0.000	0.009
	[0.010]	[0.006]	[0.010]
D.Residential Consumption (log GWh)		−0.199	−0.062
		[0.111]	[0.118]
D.Source Hydro. (log GWh/Tj)		0.002	−0.042
		[0.015]	[0.029]
D.Source Comb. Fuel (log GWh/Tj)		0.030	−0.006
		[0.022]	[0.035]
D.Imports (log GWh/Tj)		0.011	0.011
		[0.009]	[0.010]
D.Energy Distribution Loss (log GWh)		0.026	−0.022
		[0.042]	[0.049]
D.GDP (log Nat.Curr. B.)		−0.865***	−1.036***
		[0.034]	[0.047]
Population (log M.)		1.183***	1.073*
		[0.332]	[0.477]
D.Cost Comb. Oil (log Nat.Curr./TOE)			0.045
			[0.026]
D.Cost Gas (log Nat.Curr./TOE)			0.089**
			[0.029]
Year	−0.002*	−0.009***	−0.008***
	[0.001]	[0.001]	[0.002]
Constant	4.739*	15.240***	12.088***
	[1.943]	[1.723]***	[2.072]
R-squared (within)	0.028	0.717***	0.845
F-statistic	2.290	65.390***	51.100
Prob > F	0.060	0.000***	0.000
Obs.	339	309	144

Note: Standard error in square brackets.
*$p < 0.05$, **$p < 0.01$, ***$p < 0.001$
D. stands for first difference. The variable *Population* is trend stationary.
Source: Our calculations on IEA (2006).

Table 4.5 Estimation of a dynamic panel
Dependent variable is log net (of tax) price of electricity for households

	Arellando and Bond estimation			
	(A)	(B)	(C)	(D)
Vertical Integr.	−0.009	−0.012	−0.009	−0.037
	[0.007]	[0.006]	[0.006]	[0.023]
Public Ownership	0.003	0.001	0.002	0.028
	[0.007]	[0.006]	[0.006]	[0.021]
Entry Regulation	0.014*	0.021***	0.016**	0.029*
	[0.006]	[0.006]	[0.006]	[0.013]
Residential Consumption			0.241***	−0.106
(log GWh)			[0.065]	[0.077]
Source Hydro.		−0.039**	−0.036*	−0.074
(log GWh/Tj)		[0.015]	[0.014]	[0.042]
Source Comb. Fuel		0.079***	0.065**	0.015
(log GWh/Tj)		[0.023]	[0.022]	[0.036]
Imports (log GWh/Tj)		0.035**	0.032**	0.076***
		[0.012]	[0.012]	[0.014]
Energy Distribution Loss		0.185***	0.128*	0.072
(log GWh)		[0.051]	[0.052]	[0.067]
GDP (log Nat.		−0.284***	−0.310***	−0.534***
Curr. B.)		[0.024]	[0.024]	[0.052]
Population (log M.)		−0.264	−0.219	−1.183
		[0.510]	[0.502]	[0.695]
Cost Comb. Oil (log				−0.079*
Nat.Curr./TOE)				[0.036]
Cost gas (log Nat.				0.069
Curr./TOE)				[0.040]
Cost Coal (log Nat.				0.162***
Curr./TOE)				[0.029]
log net price (−1)	0.750***	0.628***	0.618***	0.357***
	[0.034]	[0.032]	[0.031]	[0.049]
Year	0.004*	0.012***	0.007*	0.035***
	[0.002]	[0.003]	[0.003]	[0.004]
Wald chi2	1349.190	1830.590	1907.750	1368.600
Sargan test chi2	406.090	411.150	411.970	151.090
Arellano–Bond test of	−4.700	−4.810	−5.000	−2.980
residual AC(1), z				
Prob > z	0.000	0.000	0.000	0.003
Arellano–Bond test of	−0.860	−0.960	−0.840	−1.030
residual AC(2), z				
Prob > z	0.387	0.337	0.403	0.303
Obs.	325	297	297	123

Notes: Standard error in brackets.
*$p < 0.05$, **$p < 0.01$, ***$p < 0.001$
Source: Our calculations on IEA (2006).

is small and that the analysis might be affected by measurement error, which could be the main reason of statistically insignificant results on regulatory indices, they should still reckon that there is no empirical evidence showing that increasing vertical disintegration and reducing the public ownership in electricity markets have a decreasing effect on energy prices for households.

When model (2) is augmented for including other stationary variables, results show that the more significant variables are population size and GDP. Also input cost variables are very relevant to explain the change of log prices, although the main drawback of the cost variables introduction in the analysis is that these variables present many missing observations, causing the sample dimension to decrease dramatically, with important effects on the significance of some other variables considered.

The goodness-of-fit measures computed for fixed-effect regressions show that when only the regulatory variables are included in the analysis only a very small proportion of total within variation[24] is explained by the model, while the explained within variation is over 70 per cent if the other variables are included showing that regulatory variables have indeed a very limited role for explaining variation across time and countries.

Finally, we estimate a dynamic model in that log prices are regressed on log prices lagged one period as well on the variables of model (1), as one might want to test whether there is any persistence in log prices. Hence, the model becomes,

$$p_{i,t} = c + p_{i,t-1}\delta + R'_{i,t}\beta + X'_{i,t}\phi + \alpha_i + t\gamma + \varepsilon_{i,t} \qquad (3)$$

Model (3) is however affected by endogeneity problems as $p_{i,t-1}$ depends upon α_i regardless of the estimation method used. The Arellano and Bond (1991) estimator is based on a generalized method of moments (GMM) to overcome the endogeneity of the lagged dependent variable exploiting a large set of possible moment conditions. Table 4.5 presents results. The Wald chi-squared goodness-of-fit statistics show that the whole vector of parameters is highly significant in all four specifications chosen. The Sargan test statistic rejects the validity of the over-identifying restrictions although it is well known that this test is over-rejecting if some heteroskedasticity of errors is present. More informative instead is the Arellano–Bond test for first- and second-order autocorrelation in the first differenced errors. The null hypothesis of this test is that there is no autocorrelation. When idiosyncratic errors are independently and identically distributed, the first-differenced errors are first-order serially autocorrelated, while serial correlation in the first-differenced error at an order higher than 1 implies that the moment condition used for GMM estimation is not valid. Results presented at bottom of Table 4.5 show that, as expected, there is strong evidence against the null hypothesis of zero autocorrelation in the first-differenced errors at order 1 but no significant evidence of serial correlation in the first-differenced errors at order 2, providing evidence of a good moment conditions for the estimation method used.

As for the single parameters, Table 4.5 shows that there is a strong autocorrelation of first differences in log prices, but that vertical integration and public ownership have an insignificant effect on log prices, while more entry regulation contributes to higher prices. As for the other variables, they show that import-dependent countries tend to have higher prices than others and that increased input costs have a cost-increasing effect on electricity prices.

6. Consumers' satisfaction with electricity prices

In the previous section we analysed whether an objective measure of an important element of consumers' welfare, such as the market price consumers pay for electricity, is influenced by the industry structure, the market entry regulation and the public ownership share in the industry. In this section consumers' subjective satisfaction is measured in the Eurobarometer data set, which collects information about approximately 1000 people in each European countries in 2000, 2002 and 2004 (for a thorough analysis of the Eurobarometer datasets concerning satisfaction with some services of general interests, see Fiorio et al., 2007).

As satisfaction to different SGI is coded with ordinal variables, analogously to Eurobarometer (2004), we dichotomize consumers' satisfaction, i.e. answers to questions about prices and quality of SGI are classified into 'satisfied' and 'not satisfied'. In particular, the consumer price satisfaction variable is recorded equal to 1 if the respondent states that the price he pays for electricity supply is fair, and is recorded equal to 0 otherwise. The consumer quality satisfaction variable is recorded equal to 1 if the respondent states that the quality of the electricity supply is very good, and is equal to 0 if the answer is fairly good, fairly bad or very bad.[25]

Table 4.6 Descriptive statistics for electricity by year and pooled sample

Satisfied with price of electricity supply

	Descriptive statistics					Countries		
Year	Obs	Mean	Std. Dev.	Min	Max	Min	Median	Max
2000	15	0.61	0.14	0.38	0.82	Portugal	Germany	Luxembourg
2002	15	0.60	0.13	0.37	0.79	Italy	Sweden	Luxembourg
2004	15	0.65	0.17	0.32	0.89	Greece	Germany	United Kingdom
all	45	0.62	0.15	0.32	0.89	Greece	Belgium	United Kingdom

Satisfied with quality of electricity supply

	Descriptive statistics					Countries		
Year	Obs	Mean	Std. Dev.	Min	Max	Min	Median	Max
2000	15	0.46	0.19	0.13	0.72	Portugal	Finland	Ireland
2002	15	0.43	0.19	0.08	0.69	Portugal	Finland	Ireland
2004	15	0.42	0.19	0.10	0.71	Portugal	United Kingdom	Denmark
All	45	0.44	0.19	0.08	0.72	Portugal	United Kingdom	Ireland

Note: Price satisfaction include very and fairly satisfied. Quality satisfaction includes only very satisfied. See note 25 for further explanation.
Source: our calculations on Eurobarometer data.

Table 4.6 shows that satisfaction with electricity supply is large across the EU. As many as 60 per cent of European consumers are satisfied with prices and over 40 per cent are very satisfied with quality. Table 4.6 reports also the countries at the bottom, in the middle and at the top of the satisfaction range, showing that some Mediterranean countries, and notably Portugal, tend to score the lowest rate of (unconditional) satisfaction.

As we do not know the exact level of individual satisfaction, S^*, for each service, we assume that satisfaction is generated by a latent variable model:

$$S^* = x\beta + e \qquad (4)$$

where $x\beta = \beta_2 x_{2c} + \ldots + \beta_k x_{kc}$ includes individual characteristics (i.e. sex, occupation) accounting for individual observed heterogeneity, time-varying country macroeconomic variables (i.e. GDP level and rate of growth) accounting for time-varying heterogeneity and a time fixed-effects variable to capture any time trend. The subscript c refers to the cluster, as it is assumed that the unobserved characteristics is $e_{ic} = \alpha_c + \varepsilon_{ic}$, where α is the cluster-specific term and ε is the idiosyncratic error term. In other words, we allow for an unobservable effect common to all households in the same country and we treat it as a fixed-effect. This is a quite general model which assumes that $Cov[e_{ic}, e_{jc}] \neq 0$, though $Cov[e_{ic}, e_{jd}] = 0$ for $c \neq d$, where i, j indicate the observation and c, d indicate the cluster, and it is reasonable in situations where the number of cluster is small relatively to the sample size. This model can be estimated directly by introducing a dummy variable for each cluster. In the present case we have a maximum of 15 clusters (EU-15 countries) with nearly 40 000 observations, and the model is estimated by cluster dummy variables model (Cameron and Trivedi, 2005), omitting the constant to avoid the dummy variable trap.

As S^* is latent, one can only observe

$$S = 1[S^* > 0]$$

where $1[\cdot]$ is equal to 1 if the argument is true and equal to zero otherwise. Assuming that ε is distributed as a standard normal, independently from x, we obtain the probit model:

$$\Pr(S = 1|x) = \Pr(S^* > 0|x) = \Pr(e > -x\beta|x) = 1 - \Phi(-x\beta) = \Phi(x\beta) \equiv p(x) \quad (5)$$

where Φ is the standard normal cumulative density function.

The partial effect of x_j on $p(x)$ depends on x through the standard normal density function, $\phi(x\beta)$, as $\partial p(x)/\partial x_j = \phi(x\beta)\beta_j$. The average partial effect (APE) for a continuous variable x_j is:

$$APE_j = \beta_j \frac{1}{n} \sum_{i=1}^{n} \phi(x^i \beta) \quad (6)$$

where n denotes the number of observations, and $x^i \beta$ the value of the linear combination of parameters and variables for the i-th observation. The APE for a dummy variable is:

$$APE_j = \frac{1}{n} \sum_{i=1}^{n} [\Phi(x^i \beta | x_j^i = 1) - \Phi(x^i \beta | x_j^i = 0)] \quad (7)$$

which avoids the problem of setting the dummy variables to their means.

All estimates to follow present results in terms of APE. As controls, x, we used a set of individual characteristics (including sex, age, marital status, age when finished education, occupation, political views, respondent's cooperation as assessed by interviewer), of country fixed-effects, year dummies, some country-level macroeconomic variables (population density, GDP per capita, GDP growth rate, employment growth rate, Gini index) and some regulatory indicators of entry regulation, public ownership, market structure and vertical integration.[26]

We include also electricity market prices for households among the independent variables of model (3), both in levels and in first difference and the consumer price index (CPI) to test whether subjective satisfaction depends on actual prices and whether the relationship between subjective satisfaction and regulatory variables is at all driven by the relationship between regulatory variables and market prices of electricity. In Table 4.7 marginal effects for price satisfaction are reported. In column (A) only market prices and CPI are included to test whether consumers' satisfaction is somehow related to prices. Estimates show that with this specification the level of market prices have the expected sign but no statistical significance. In column (B) regulatory variables are introduced leaving out market prices and CPI. It shows that regulatory variables are all highly significant. As the probit model is a non-linear model, the interpretation of the marginal effects is different from the case of linear regression models and in particular they closely represent the marginal change on the dependent variable (probability of satisfaction) due to a marginal change of an independent variable only for an infinitesimal change in the dependent variable. So, if the public ownership indicator increased by an infinitesimal amount (as measured in the REGREF data set), the probability of consumer's satisfaction increases by an average of 2.6 per cent times the change of the dependent variable. If vertical integration increases and entry is more regulated, each by a marginal amount, probability of satisfaction increases by about 1.3 per cent and 0.21 per cent. Table 4.7 also shows that country fixed-effects are the most relevant factors determining the consumers' satisfaction, as they also include other country-specific omitted variables. In column (C) we introduce at the same time the regulatory variables and the market price for electricity in national currency per kilowatt/ hour, in levels and first difference, and the CPI. If prices and regulatory variables were highly correlated, this would cause a problem of collinearity which increases the standard errors of estimates. Instead, column (C) shows that regulatory variables do not lose much of their statistical significance and are robust in their sign and magnitude, while price levels are now statistically significant, with large coefficient and the expected sign, suggesting that results of column (A) were most likely affected by omitted variables

Table 4.7 Average partial effect of consumers' satisfaction with electricity prices

	Electricity: Price (A)	Electricity: Price (B)	Electricity: Price (C)
Regulatory variables			
Public Ownership: Ele		0.026***	0.027***
Vertical Integration: Ele		0.013***	0.012*
Entry Regulation: Ele		0.021***	0.040***
Price variables			
Price (in local currency per Wh)	−0.444		−1.334**
First difference of Price (loc. curr. per Wh)	0.000		0.000
CPI	−0.008**		0.012***
Year dummies			
year 2002	−0.001	−0.003	0.040
year 2004	0.148***	0.120***	0.080**
Individual characteristics	Yes	Yes	Yes
Macroeconomic controls	Yes	Yes	Yes
Country fixed-effects			
Denmark	0.390***	−0.540**	0.389***
Germany	−0.332	−0.553**	−0.553**
Greece	0.313*	−0.589***	−0.622***
Italy	−0.269	−0.636***	−0.636***
Spain	0.356**	−0.364	−0.601***
France	0.316	−0.524**	−0.585***
Ireland	0.383***	0.072	−0.585***
Netherlands	−0.593***	−0.593***	−0.593***
Portugal	0.296	−0.605***	−0.618***
Great Britain	−0.124	−0.504***	−0.505***
Finland	0.379***	0.281**	−0.509**
Austria	0.330**	−0.433	−0.600***
Observations	36007	36007	36007
log-likelihood	−22274.764	−22239.426	−22232.140

Note: Robust p values in brackets (using the Huber/White/sandwich estimator of variance).
* significant at 10%; ** significant at 5%; *** significant at 1%
Source: Our calculations on Eurobarometer, IEA (2006) and REGREF data.

bias. The negative sign of the estimated price level coefficient shows that a marginal increase of prices would have a very strong impact on consumers' satisfaction reducing their satisfaction by over 130 per cent (times the marginal increase of price). The positive sign of the consumer price index (CPI) variable in column (C) shows that if consumer price dynamic is strong, the satisfaction of electricity prices tends to be higher. As the constant is missing, the country fixed-effects are also measuring the country-specific omitted variable. However, a rapid inspection of regression results shows that for many countries there is no significant difference in the average conditional level of satisfaction and, in particular, that consumers living in the UK, the country where electricity services have been liberalized the most and where privatization of the electricity services is now complete, are not more likely to be satisfied than most other countries. Finally, the time dummies coefficients in Table 4.7 show that across the three years considered, the trend is towards a significant increase of consumers' satisfaction.

Table 4.8 shows the results of the estimation of similar models where the dependent variable is now the probability of consumers' (high) satisfaction with electricity supply quality. Results show that the quality satisfaction is more variable and that fewer variables are significant. In particular, the prices coefficients are positive, showing that higher prices are correlated with perception of higher quality, and this correlation is also highly statistically significant. Among regulatory variables, only public ownership is always significant at least at the 5 per cent level and its APE is robust even if prices are introduced. Coefficients of entry regulation are significant at the 1 per cent level only if no price is introduced, but its APE and statistical significance reduce dramatically if prices are included. However, the estimated coefficients in the regression are always positive showing that higher public ownership, more integrated industry and entry regulation have a positive effect on consumers' satisfaction with electricity service quality. The overall trend is also improving across years.

7. Summary and conclusions

Our empirical results do not support the view that a unique reform paradigm is dominant in terms of welfare changes across the EU-15 Member States. We summarize below the main findings.

First, we have tested the impact of regulatory indicators on retail electricity prices. Different empirical models show that:

a) The impact of regulation (as measured by the REGREF indicators) on the change of log electricity prices is rather small, not statistically significant, and without adding other variables has very limited explanatory power

Table 4.8 Average partial effects of consumers' satisfaction with electricity quality

	Electricity: Quality (A)	Electricity: Quality (B)	Electricity: Quality (C)
Regulatory variables			
Public Ownership: Ele		0.019***	0.017***
Vertical Integration: Ele		0.006	0.001
Entry Regulation: Ele		0.024***	0.002
Price variables			
Price (in local currency per Wh)	1.618***		2.039***
First difference of Price (loc. curr. per Wh)	0.000		0.000
CPI	−0.020***		−0.015***
Year dummies			
year 2002	0.022	0.038	−0.010
year 2004	0.148***	0.043*	0.090**
Individual characteristics	Yes	Yes	Yes
Macroeconomic controls	Yes	Yes	Yes
Country fixed-effects			
Denmark	0.006	0.638***	−0.348***
Germany	0.585***	0.348	0.120
Greece	0.632***	0.045	0.631***
Italy	0.537***	0.163	0.235
Spain	0.581***	0.342	0.580***
France	0.609***	0.455	0.608***
Ireland	0.650***	0.634***	0.650***
Netherlands	0.648***	0.372	−0.352***
Portugal	0.622***	−0.113	0.598***
Great Britain	0.648***	0.542	0.529
Finland	0.644***	0.606***	0.644***
Austria	0.651***	0.628***	0.651***
Observations	35651	35651	35651
log-likelihood	−21438.289	−21448.080	−21432.691

Note: Robust p values in brackets (using the Huber/White/sandwich estimator of variance)
* significant at 10%; ** significant at 5%; *** significant at 1%.
Source: Our calculations on Eurobarometer, IEA (2006) and REGREF data.

b) When we include in the model other stationary variables, the more significant variables are population size and GDP, pointing to exogenous determinants of prices
c) Input costs contribute to explain electricity prices, with the expected sign (e.g. hydro power decreases the final price)

d) When we estimate a dynamic model (log prices regressed on previous period as explanatory variable) we find high persistence of log prices
e) In this context both vertical integration and public ownership have an insignificant effect on log prices, while more entry regulation contributes to higher prices. This is the only evidence we have found that liberalization can have a beneficial, albeit modest, impact on price dynamics
f) As one may expect, import-dependent countries tend to have higher prices than others and increased input costs have a cost-increasing effect on electricity prices.

Second, we have considered consumers' preferences and perceptions. We have found that:

g) When we test whether electricity prices and a general inflation index influence consumers' satisfaction on prices they pay, without the regulatory variables and other controls, the level of market prices has the expected sign but no statistical significance
h) When regulatory variables are introduced, leaving out market prices and CPI, the regulatory variables are all highly significant, but the sign of the coefficient is opposite to what would be expected by the paradigm. A marginal increase of public ownership increases consumers' satisfaction, and the same happens with vertical integration and entry regulation
i) In general country fixed-effects are the most relevant factors determining the consumers' satisfaction, pointing to omitted variables
j) When we use electricity prices and regulatory variables jointly as explanatory variables of consumers' satisfaction (in order to capture other effects that can be concealed by aggregate price data), regulatory variables coefficients are still statistically significant, robust in their sign and magnitude, while price levels are now statistically significant, with large coefficients and the expected sign
k) After controlling for the individual characteristics in the Eurobarometer samples, for many countries there is no significant difference in the average conditional level of satisfaction. British consumers are not more likely to be satisfied than most of other 'unreformed' or less reformed countries.
l) The determinants of consumers' satisfaction with electricity service quality are more elusive. Higher prices seem to be associated with a perception of higher quality. Again we find unexpected results

from the viewpoint of the paradigm: higher public ownership, more integrated industry and entry regulation have a positive effect on consumers' satisfaction with electricity service quality.

In a nutshell: if we consider both aggregate country data on prices and micro-data on consumers' satisfaction, public ownership tends either to be statistically associated or associated with decreasing prices and higher consumers' satisfaction; unbundling tends to increase prices (or to be not significant) and to lower consumers' satisfaction; entry barriers, as may be expected, perhaps do decrease prices but consumers are not convinced. Input costs, country-specific features, and other country controls, have in general higher explanatory power than regulatory variables. Results on quality are less easy to interpret, but again do not provide a clear support to the reform paradigm.

How can we interpret these findings? If one strongly believes that the paradigm must work, one may conjecture either that the data do not capture adequately the benefits of reforms, or that the indices supplied by the OECD regulatory reform database do not capture all the subtle dynamics involved. One may also think that in some countries it is too early to draw conclusions.

While we cannot entirely dismiss these three sets of possible objections, we stress that our estimations are based on databases produced by the same international organizations that are most supportive of reforms, that the econometric tests we perform are quite easily replicable with a different database. Moreover, the time span of observation is reasonably long compared with many other industry reforms.

We do not conclude that the paradigm is under any circumstances not keeping its promises. We would rather suggest that probably some of its assumptions are perhaps too strong or oversimplified and empirical analysis shows that more flexibility and realism is needed.

As an indication for further research, we reconsider in turn the three cornerstones of the paradigm: privatization, unbundling and liberalization, and why in some cases empirical analysis may lead to counterintuitive results.

Privatization is certainly in the weakest logical connection with the other two sets of reforms. The Nordic countries show that highly competitive national markets and a regionally integrated transboundary market are well supported by an industry structure where public ownership plays a remarkable role. The fact that generators are often owned by municipalities can be seen as an intrinsic constraint to anti-competitive mergers and acquisitions, which are often motivated not by economies of scale in production but by the desire to influence prices. The same reasoning may apply to a public sector owned firm exposed to competition. In contrast, UK, Belgium, and Spain industries are fully under private ownership, but their performances are so different, from the very good to the very bad, that it is difficult to attribute a clear role to privatization per se in these countries, after having controlled for other factors. The same applies for the entirely publicly owned industries in Ireland, France, and Greece. In fact best and worst performers are in both groups.

The main case for privatization, as mentioned in the opening section, is an efficiency argument. There are some factual problems in the electricity industry with this argument, however, that can explain our rather counterintuitive findings. First, in terms of productive efficiency, the electricity industry is based on rather well established technologies, that are more or less common knowledge. The quantity/price combinations available to generators are largely dictated by priority/cost merit rules and other rationing mechanisms that greatly reduce managerial discretion. Thus, the extent of asymmetric information between stakeholders is probably reduced in the electricity industry compared with other industries. Hence, for a well-motivated public sector management, and a benevolent government, the possibility for the former to earn rents based on a principal–agent mechanism should not be exaggerated.

Second, public sector managers can collude with policy makers and trade unions to allow for labour hoarding, gold-plating of plants, and distributing rents. The extent of such behaviours is however rather easy to be detected, and a government that is under hard budget constraints, or that is exposed to scrutiny over corruption because of the democratic system of checks and balances, will refrain from allowing too much rent-seeking in a highly technical industry. An example is the USA where a substantial fraction of power generation is controlled by the local public sector and where, under strong regulation, the differences in productivity are not significant (Pollitt, 1995).

On the other side, one should not exaggerate the efficiency case for private ownership. Major players are often, after all, large firms managed by boards of directors who respond to coalitions of shareholders. The latter, in turn, are often quite at arm's-length distance from them, as happens when they are financial investors. Those investors may well have objectives that cannot be described as outright profit maximization, for example because they are pension funds looking more for stable returns than for taking the risks usually associated with innovation. A close scrutiny of who actually are the 'private' owners of the privatized electricity industry may show that in many cases they do not need to be described as aggressive profit maximizers. Thus, the interplay between the objectives of top executives and shareholders in the privatized electrical utilities should be viewed with a dose of sober realism, rather than assuming efficient behaviour.

Moreover, when we focus on price performance, the role of ownership in the electricity industry is even less clear-cut than the new paradigm would imply. Basically, consumers pay a price that can be broken down to a number of components: indirect taxes, retailing margins, costs for transmission and distribution, costs for generation. Clearly, indirect taxes have nothing to do with the other components because governments decide them. If generators optimize over the energy mix, generation costs are largely dominated by the prices of inputs and the technical efficiency of plants. The picture here is much more blurred than one would expect because indeed the mix of energy sources available is the result of a number of policy decisions, most of them external to the owner of the generating plant. Examples are when Italy rejected the building of nuclear power plants after a referendum; the existence of subsidies to some kinds of energy sources in Germany; the complex licensing of imports of gas from Russia; and the nuclear national strategy in France, etc.

Eventually, the overall mark-up over costs under private ownership is not necessarily lower than under public ownership. The overall mark-up over costs depends probably more upon the interplay of regulation and competition than upon ownership per se. Standard public economics theory would say that the mark-up over costs should be rather lower under public than under private ownership, if regulation or competition does not perfectly substitute for long-run marginal cost pricing rules in the public sector (and, of course, if these rules were more or less applied before reform) again after controlling for technological constraints.

Actually our empirical findings seem consistent with the latter interpretation, because they show, at least for the countries and the years we have considered, that both objective data on prices and consumers' perceptions converge in predicting lower prices under public ownership, or neutrality of ownership on prices.

Thus, is there something going wrong with the reforms? This is difficult to say without a detailed analysis, but we suggest two examples that may question the universal validity of the new paradigm.

There is some evidence that the general case for unbundling was too strong. First, when regulators allowed to do it, electric utilities quickly tried to re-integrate vertically and the resulting combinations in some countries were more cost-effective and more resilient to shocks than their competitors. These trends reinforced the view in the business that the costs of vertical disintegration should be assessed along with its benefits and there seems now to be a shift of opinions against ownership separation of networks from generation (the EC however insists on this reform).

According to Glachant and Leveque:

> The industrial reference model for electricity reforms completely changed between 1995 and 2001. It has shifted from a preference for structures that are vertically disintegrated between generation, trading, and sale to final consumers toward a preference for vertical reintegration. Bankers and financiers have finally joined companies with stockholders and managers and concluded that vertical integration is the best protection against volatility and the cyclical nature of markets. Nowadays, most national and European energy markets involve firms that are vertically integrated. (2005, p.10)

If this view prevails, and at the same time horizontal mergers and acquisitions processes go ahead as we can observe for several years in the EU, clearly the foundations of the reform paradigm are even more questioned. Competition by and large should happen only by regulating third-party access to infrastructures owned by some of the players. It is not evident that it is more difficult to achieve this if the owner is a private firm than with a public one. Under a TPA regime a private unbundled or re-bundled company should constrain its generation capacity to allow for a competitor to use its infrastructure. One may think that it would be much easier to do so with a publicly owned, vertically integrated, *but non monopolistic,* public sector firm, operating under strict incentive regulation, and with clear public service obligations.

This brings us to the last issue: some policy implications of our findings. If the overarching goal of the European Commission is to stimulate the creation of an internal market, that is, a wider trade of electricity across the EU, the focus should probably shift from a dubious unique reform paradigm, to be implemented by each country, to a more substantive EU-wide energy policy initiative. The true concern should not be whether EDF is publicly owned, or if it is fully unbundled, but whether the French consumers (and firms) can buy electricity from competitors, including from those generating electricity abroad, if the latter are ready to supply it at tariffs lower than to EDF.

The constraints that prevent this fundamental consumer protection mechanism from electricity monopoly power in the EU are related to the lack of physical investment in trans-European energy networks (the TEN-E), and in establishing the rules of the game and the institutional arrangements for allowing international trade of electricity at least at regional level, as in the Nordic example. After all, the true success story of the European project since the 1960s, lies in opening borders to trade not in dictating detailed market structures deemed to be the most efficient and competitive. Let us open the border to electricity trade and we shall then see which are the best industry patterns (not necessarily the same in Greece and Sweden, however). The fact that in our empirical models imports do not concur to lowering prices is probably additional evidence that currently interconnection and trade are often a residual mechanism for most countries, capturing only marginal demand with higher willingness-to-pay.

Ironically, for the time being, a strong policy option in favour of international trade of electricity may imply that in an integrated European market, some state-owned, perhaps vertically integrated, companies will be more competitive than some national privately-owned ones, currently de facto protected by the lack of physical interconnections, incomplete institutional arrangements for trade or by other forms of sheltering regulations. It is sufficient to look to the dispersion of electricity tariffs in the EU to understand that, after all, the key policy step forward should be to the opening of national systems, whatever they are, to the challenge of competition from abroad. There is no point in reshuffling ownership structures and other aspects of the industry in a mainly domestic perspective, seeking structural uniformity. If the overarching goal is to offer consumers the best quality and price of service in an integrated Europe, international market opening seems to be by far more important than imposing privatization, unbundling and domestic liberalization, when those reforms do not deliver the most socially efficient outcomes given the country-specific features.

The neutrality of the founding fathers of the European project as regards public ownership was wise because the ownership structure of essential services in any society is probably more about the desired balance of economic and political powers than about efficiency and competition. Economists sometimes tend to misunderstand what ownership actually is because they are inclined to apply a very simplified model of the incentives it provides and to disregard political, social and subtle legal aspects of property rights. The balance of public and private property rights should be left to governments and citizens to consider and to decide, when the relative merits in terms of productive efficiency or of market behaviour are uncertain, or limited. If municipalities want to produce their power in Nordic countries, or France to centrally manage its nuclear industry, this is something that should not be questioned by the EC or by any other international organization. Moreover, unbundling network industries under certain circumstances can work; in other cases it may be too costly. Liberalization on a national basis may be less useful than expected if the domestic producers, whatever their ownership, are sheltered by international competition. However, a European consumer, or a representative body acting on his behalf, should be given the right and the opportunity to buy electricity from whoever offers it at the lowest prices, for a given quality, and under secure long-term supply arrangements, in a sustainable environment. Thus an integrated EU energy strategy, with its incentives and disincentive mechanisms, should replace the obsession with dictating a uniform industry reform paradigm.

Notes

1. First European Electricity Directive n. 92, 1996.
2. Second European Electricity Directive n. 54, 2003.
3. The data reported in section 3.2 and 3.3 are from Eurostat.
4. Serralles (2006) and Al-Sunaidy and Green (2006).
5. Commission of European Community (2006).
6. Transmission System Operator.
7. Third-party access.
8. Commission of European Community (2006) and (2007).
9. Ferrari and Giulietti (2004) and Commission of European Community (2006).
10. Commission of European Community (2006).
11. Norway was the first Nordic country to liberalize its electricity market in 1991, following the British model in most respects, but without privatization. The state now still plays a leading role through local distributors and in some cases also public producers.
12. Thomas (2006) and Commission of European Community (2006).
13. Thomas (2006) and Commission of European Community (2006).
14. Thomas (2006) and Commission of European Community (2006).
15. Serrales (2006) and Commission of European Community (2007).
16. Serrales (2006) and Commission of European Community (2007).
17. Serrales (2006) and Commission of European Community (2007).
18. Ferrari and Giulietti (2004) and Commission of the European Community (2007).
19. Reisch and Micklitz (2007) and Commission of the European Community (2007).
20. Thomas (2006) and Commission of the European Community (2007).
21. Thomas (2006) and Commission of the European Community (2007).
22. Thomas (2006) and Commission of the European Community (2007).

23. The Dickey–Fuller test is not presented here but may be obtained from the authors.
24. As the fixed-effect estimator is chosen to explain the within variation as well as possible, it maximizes the 'within R^2, which is given by $R^2_{within}(\hat{\beta}_{FE}) = corr^2(\hat{p}_{it}^{FE} - \hat{P}_i^{FE}, \hat{p}_{it} - \bar{p}_i)$, where $\hat{y}_{it}^{FE} - \hat{y}_i^{FE} = (x_{it} - \bar{x}_i)'\hat{\beta}_{FE}$ and $corr^2$ is the squared correlation coefficient.'
25. Some readers might be puzzled by the fact that we include among the non-satisfied those who declared that quality of SGI is fairly good; however, this is due simply to increased variability. In fact, only about 5% of consumers across services rate quality of SGI as fairly or very bad.
26. It would be meaningful to include a variable of household income but it is not available for the whole period considered. However, some of the variables included, such as education, occupation and age of the respondent are likely to be highly correlated with income.

References

Al-Sunaidy, A. and R. Green (2006) 'Electricity deregulation in OECD countries', *Energy*, vol. 31: 769–87.
Arellano, M. and S. Bond (1991) 'Some tests of specification for panel data: Monte Carlo evidence and an application to employment equations', *Review of Economics Studies*, 58: 277–94.
Cameron, A.C. and P.K. Trivedi (2005) *Microeconometrics, Methods and Applications*, Cambridge University Press: Cambridge.
Commission of the European Community (2005a) – General Direction for Energy and Transport '*Report on Progress in Creating the Internal Gas and Electricity Market*', Bruxelles, 2005.
Commission of the European Community (2005b) – General Direction for Energy and Transport '*Report on Progress in Creating the Internal Gas and Electricity Market*' – Technical annex', Bruxelles, 2005.
Commission of the European Community (2006) – General Direction for Energy and Transport '*Study on Unbundling of Electricity and Gas Transmission and Distributors Operators*' – Technical annex', Bruxelles, 2006.
Commission of the European Community (2007) – General Direction for Energy and Transport '*Prospect for the Internal Gas and Electricity Market: Implementation Report*', Bruxelles, 2007.
Conway, P. and G. Nicoletti (2006) 'Product market regulation in non-manufacturing sectors in OECD countries: measurement and highlights', OECD Economics Department Working Paper.
Copenhagen Economics (2005) *Market Opening in Network Industries*, Report by Copenhagen Economics on behalf of DG Internal Market, EU Commission, downloadable from http://www.copenhageneconomics.com/publications/
Eurobarometer (2006) 'The Constitutional Treaty, Economic Challenges, Vocational Training, Information Technology at Work, Environmental Issues, and Services of General Interest', Report No. 62.1.
Ferrari, A. and M. Giulietti, (2004) 'Competition in electricity markets: international experience and the case of Italy', *Utilities Policy*, Vol. 13: 247–55.

Fiorio, C.V., M. Florio, S. Salini and P.A. Ferrari (2007) 'Consumers' attitudes on services of general interest in the EU: Accessibility, price and quality 2000–2004', *Nota di Lavoro FEEM*, N. 2.

Florio, M. (2007), 'Electricity prices as signals for the evaluation of reforms: an empirical analysis of four European countries', *International Review of Applied Economics*, Vol. 21 (1): 21–7.

Glachant J.-M. and F. Leveque (2005) 'Electricity internal market in the European Union: What to do next?', paper presented at the conference 'Implementing the internal market: Proposal and timetables, Brussels, 9 September 2005, available at www.sessa.eu.com

International Energy Agency (2006) *Monthly Electricity Statistics*.

International Energy Agency (2006) *Electricity Information*.

Lampietti J.A., S.G. Banerjee, A. Branczik (2007) *People and Power. Electricity Sector Reforms and the Poor in Europe and Central Asia*, World Bank, Washington D.C.

Levi Faur, D. (2003) 'The politics of liberalization: privatization and regulation for competition in Europe's and Latin America's telecoms and electricity industries', *European Journal of Political Research*, Vol. 42: 705–40.

Jamasb, T. and M. Pollitt (2005) '*Electricity Market Reform in the European Union. Review of Progress toward Liberalization and Integration*', CEEPR wp 3-2005.

Millward, R. (2005) *Private and Public Enterprise in Europe. Energy, Telecommunications and Transport 1830–1990*, Cambridge University Press: Cambridge.

Newbery, D. and Pollitt, M. (1997) 'Restructuring and privatization of CEGB – Was it worth it?', *Journal of Industrial Economics*, 45 (3): 269–304.

Pollitt, M. (1995) *Ownership and Performance in Electric Utilities: The International Evidence on Privatization and Efficiency*, Oxford University Press: Oxford.

Reisch L.A. and Micklitz, H.W. (2007) 'Consumers and deregulation of the electricity market in Germany', *Journal of Consumer Policy*, Vol. 29: 329–415.

Serralles, R.J. (2006) 'Electric restructuring in the European Union: integration, subsidiarity and the challenge of harmonization', *Energy Policy*, Vol. 34: 2452–551.

Thomas, S.D. (2006) 'Electricity industry reforms in smaller European countries and the Nordic experience', *Energy*, Vol. 31: 788–801.

5
Privatization and Deregulation of the European Electricity Sector

Catalina Gálvez, Ana González, and Roberto Velasco

Universidad del Pais Vasco

Abstract

The electricity sector liberalization was extended into Europe with the publication of the Directive 96/92/EC on common norms for the internal electricity market, which encouraged Member States to foster competition with the aim of creating a single European market. However, diversity of interests has induced the European countries to behave evasively about those purposes in some essential aspects, which has so far prevented its transformation into tangible results.

In order to make a profound study of those aspects decisive for the attainment of the objectives fixed by European institutions this work tries to determine, through several indicators, the level of fulfilment of the Community Directives in the different countries, as well as to confirm whether competition in generation and supply has actually increased. In this regard, the analysis of privatization and concentration processes that have taken place in parallel to liberalization in some of the most relevant countries of the EU shows the appearance of big conglomerates with predominant status in the respective domestic markets and in the European context also. Such a situation favours neither competition nor the subsequent generalized reduction of the price of electric energy for consumers.

In consequence, this chapter finds that the merger and acquisition process has proved to be one of the main obstacles, together with regulatory unevenness and the different pace of countries to adapt themselves to the Community Directives in the progress towards a real single internal electricity market.

Keywords: privatization, deregulation, liberalization, concentration, internal electricity market, competition.

Journal of Economic Literature classification: L1, L33, M48, N7, Q4

1. Introduction

The special characteristics of electric energy have important implications for the performance and design of markets. Among these characteristics, the following can be highlighted. On the one hand, the electricity sector includes different activities: (1) the production or generation of electricity, (2) the transport of electricity on high voltage levels (transmission), (3) its transportation on low voltage levels (distribution), and (4) the marketing of electricity to final consumers (supply). The electricity sector is a network industry, and it is considered a natural monopoly that can not be duplicated.

On the other hand, electricity cannot be stored once produced, which requires a balance between production and consumption together with certain surplus capacity in order to cover the peaks of the load curve of a demand that fluctuates significantly during the day and seasonally. Also the price elasticity demand is very low, especially over the short term, since there are no substitutes and it is fundamental for other production activities and the welfare of society. At the same time electricity generation requires relatively large capital investment and long periods to build new power plants.

Electricity can be produced in many ways, using a variety of fuels and applying different technologies, resulting in different cost structures (nuclear, hydro, coal, gas, renewable, etc.). These different technologies costs have important implications for the price formation on electricity market.

The concept of a competitive market, applied to the electricity sector, means getting a balance with enough suppliers, which is, from the consumer point of view, an optimum situation both regarding prices and security of supply. However, the European reality, as we will see later, is proving difficult in some essential aspects, especially with regard to the level of competition reached in its internal market as a consequence of the mergers and acquisitions taking place in the European industry with the consequent increase of its level of concentration. This situation has brought about the appearance of large conglomerates with a predominant position in their national markets and in the European context. They are known as 'national champions'. The purpose of this study is the analysis of liberalization and privatization processes in the main European countries, from the viewpoint of both producers and consumers of electricity.

162 *Critical Essays on the Privatization Experience*

With this aim, and after a brief introduction of the sector, we offer a historic perspective of the privatization and deregulation processes experienced by the world electricity sector and later we will focus on the process taking place in Europe. After reviewing the Community Directives that have regulated the creation of a single electricity market we have tried to determine, through several indicators, the level of fulfilment of such directives in the different countries and to ascertain if competition has increased.

Finally, we analyse the processes of concentration and privatization of companies that have helped the electricity sector liberalization process in the United Kingdom, Spain, Germany, Italy and France, where the most important European electricity companies are located.

2. Brief presentation of the European electricity industry

Figure 5.1 shows electricity production in EU-25, in terms of its generation sources in 1990 and in 2004. In both years analysed the main source of electricity is conventional thermal energy, followed by nuclear energy. Both sources together accounted for 87 per cent of the total electricity production in 2004. The rest goes between hydro, wind and geothermal energy and others.

Figure 5.2 offers an overall view of the final consumption of electricity in the EU-25 context. Once again we take 1990 and 2004 as reference years. The analysis of the data clearly shows that, in the period under study, there has been a considerable increase in the relative electricity consumption by household and service sector use and relative decline in that of industry. Thus, whereas the industrial sector's share of consumption of electricity has dropped by 3.8 per cent, household use and service sector have gained 4 per cent. This does not mean the industrial sector's consumption has decreased in absolute terms but household and service consumption has increased much faster.

Table 5.1 collects data relative to production, consumption and net imports of electricity. In the EU-25 framework, the bulk of electricity production has increased 33.5 per cent between 1990 and 2004. If we bear in mind production levels in 2004, Germany was the main producer among the countries which are the object of this study, followed by France. At a certain distance we find the United Kingdom and Italy. Regarding variation it is important to highlight the Spanish electricity sector, which has enjoyed the highest growth in the years considered, with an

1990

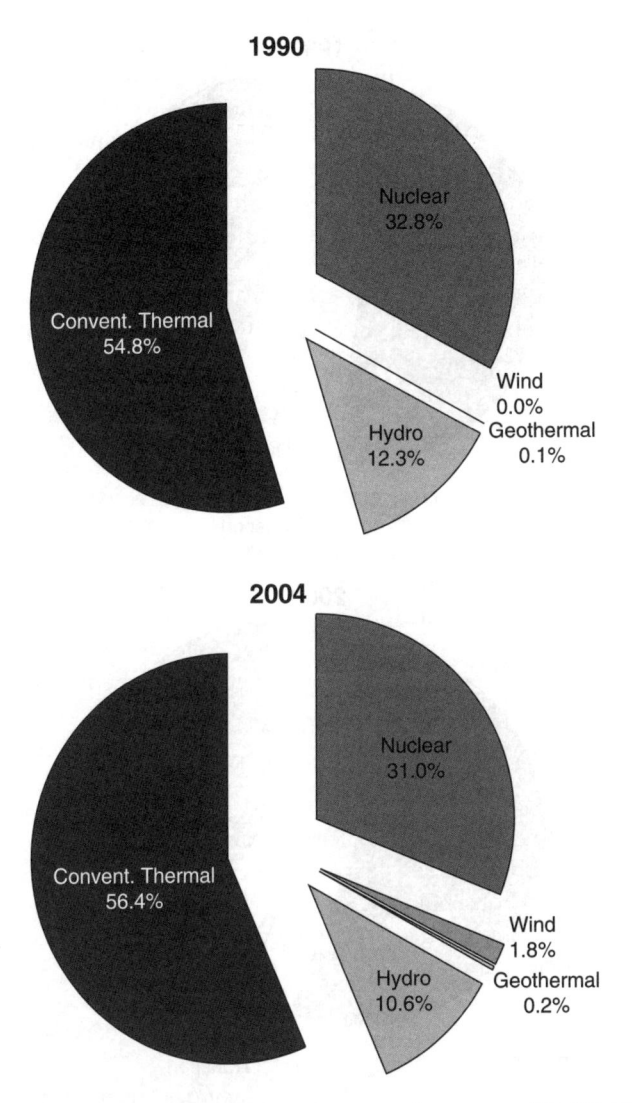

- Convent. Thermal 54.8%
- Nuclear 32.8%
- Hydro 12.3%
- Wind 0.0%
- Geothermal 0.1%

2004

- Convent. Thermal 56.4%
- Nuclear 31.0%
- Hydro 10.6%
- Wind 1.8%
- Geothermal 0.2%

Figure 5.1 Electricity production as per its generation sources (%), EU- 25 (1990, 2004)
Source: Eurostat. Our estimation.

84.3 per cent increase in its electricity production. At the other extreme we find the United Kingdom with the lowest growth rate of 24.1 per cent. The last block of data of the table makes reference to the commercial balance of electric energy. All the countries considered are net importers

1990

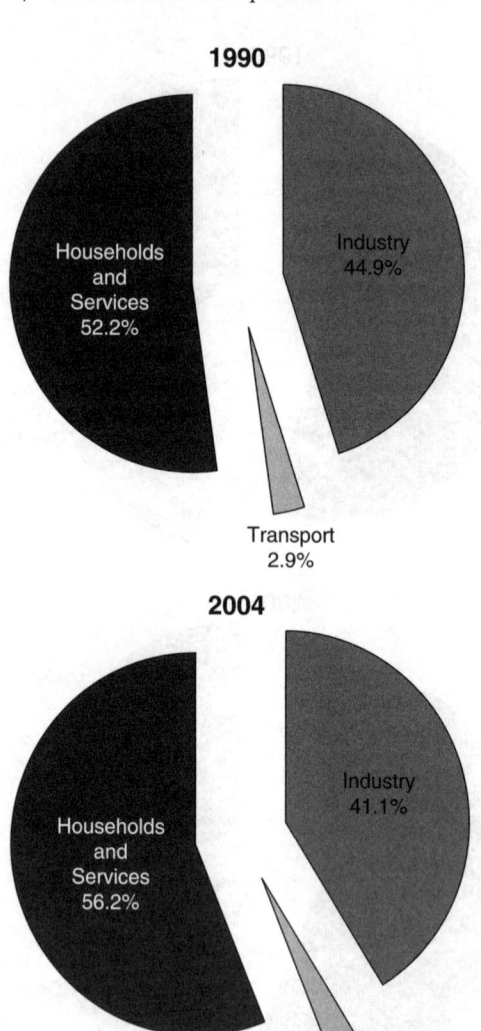

2004

Figure 5.2 Electricity consumption as per activity, EU-25 (1990, 2004)
Source: Eurostat. Our estimation.

of electricity, except France. Spain and Germany reached a favourable balance in the electricity foreign market in 2004, but their previous evolution shows the exceptional nature of the present data (see Eurostat 1990–2004).

Table 5.1 Production, consumption and net imports of electricity, 1990, 2004

Country	Production Mtoes 1990	Production Mtoes 2004	Production Change % 1990/04	Consumption GWH 1990	Consumption GWH 2004	Consumption Change % 1990/04	Net Import GWH 1990	Net Import GWH 2004	Net Import Change % 1990/04
EU-25	2 380 776	3 179 132	33.5	2 051 876	2 651 682	29.2	25 363	−422	−101.7
EU-15	2 061 619	2 820 466	36.8	1 813 456	2 405 313	32.6	27 134	26 654	−1.8
Germany	453 591	606 636	33.7	446 489	513 327	14.9	789	−2 621	−423.2
Spain	151 838	279 953	84.3	125 799	230 669	83.6	−420	−3 028	621.0
France	420 744	572 241	36.0	301 912	415 880	37.7	−45 438	−62 040	36.5
Italy	216 878	303 322	39.8	214 084	295 042	37.8	34 655	45 635	31.7
UK	318 963	395 853	24.1	274 433	340 042	23.9	11 943	7 490	−37.3

Note: Mtoes: Million tons of oil equivalents.
Source: Eurostat. Our estimation.

3. Background of privatization in the electricity sector

From the 1990s the electricity sector has witnessed an intense deregulation and privatization process although, as Beder (2005) indicates, it has been a change in the regulation system rather than the disappearance of such regulation: first, by protecting its functioning as a public service and later aiming at the smooth running of the market. Therefore, we consider the term 'privatization' describes what happened much more accurately. As Fabra (2004) observed:

> the concepts liberalization and regulation are not antithetical but complementary. It is not possible to liberalize without regulation since liberalization without regulation would lead to the development of power market positions and collusive attempts. The opposite of liberalization is discretionary intervention, only acceptable under extreme circumstances (p. 51)

During the 1970s the main western economies started to apply principles of economic policies with a strong neo-classical background. This new economic model was known as neo-liberal-monetarist system and one of its basic principles was the reduction of state intervention in the economy. This meant policies for cutbacks in public expenditure, privatizations to restructure public accounts and, as we said before, deregulation of certain economic activities, among them trading activities.

One of the first countries to adopt reform in the market-oriented electricity sector was Chile. This government divided public utilities in different parts and later sold them in the 1980s.[1] Developed countries also took steps to privatize the electricity sector, Australia and the United Kingdom being the countries with the greatest significance processes at the beginning of the 1990s. This privatization wave came together with the arrival of foreign companies, mainly through mergers and acquisitions; both in Australia and in the United Kingdom, the biggest foreign investors were US companies (EIA, 1996).

In the case of United States, deregulation was mainly promoted by the interests of companies: on the one hand, large electricity consumers, who wanted to cut down on their cost by means of agreements with suppliers, and on the other, private energy companies, which tried to obtain earnings by entering the electricity business (Beder, 2005). The first state to undertake deregulation was California in 1996.

In Latin America, Asia and Africa, privatizations were introduced by pressures from the World Bank and the IMF (Beder, 2005).[2] Between 1988 and 1993 about 2700 state-owned companies were transferred to the private sector in 95 countries (EIA, 1996).[3] In these developing countries, privatization has included the construction of new generation capacities and transport networks. In Latin America privatization of electricity generation companies has been general, with the exception of Cuba, and has been closely related to the privatization of natural gas exploitation. Many international oil companies (mainly those with an important gas production and transport business) have integrated vertically, including electricity generation.

The process described has not been devoid of errors and conflicts. The Californian supply crisis in 2000,[4] Enron's bankruptcy, the malfunctioning of 'the British pool' or the 'national champions policy' are mere expressions of the malfunctioning of the world electricity market that concerns experts, policy makers and consumers.

4. Regulation of electricity sector in Europe

Liberalization of the electricity sector began in Europe in 1996 with the approval of the Directive 96/92/EC on common norms for the internal market for electricity,[5] which encouraged countries to open their markets to competition with the purpose of creating a single electricity market in Europe, which is still far from becoming a reality.

This Directive introduced the distinction between regulated and competitive activities. Transmission and distribution activities are considered to be natural monopolies: there are economies of scale in social costs so they remain regulated. However, generation and supply activities have been progressively opened to competition.

Regarding transmission and distribution, it proposes the creation of two figures: the transmission system operator and the distribution system operator, which are responsible for coordination and management of transmission and distribution networks, respectively. On the other hand, this Directive demands the existence, at least, of accounting separation between generation, transmission and distribution activities. Moreover, Member States are obliged to open their wholesale markets to large consumers and distributors.

The internal electricity market has been progressively implemented in the different Member States, with stronger impulses since 1999–2000. All through this time the concept of a single market has been spreading through a complex legal framework, with the transposition to a national level of the different community Directives, by reaching agreements and with the creation and improvement of physical infrastructures.

In 2003 the second Directive on electricity was published,[6] providing the key legislation to establish the Internal Electricity Market. This Directive aimed at total liberalization of the electricity market. It required that large electricity consumers were free to choose electricity supplier ('eligible') by July 2004. This would be followed by the opening of the electricity markets for all consumers by July 2007. On the other hand, in order to limit the risk of discriminatory access to networks, the second Directive requires the existence of legal unbundling between transmission operators, distribution operators and the rest of the sector agents. At the same time Member States are obliged to introduce 'regulated third party access' regimes, which guarantee equitable access of new suppliers to the network based on public tariffs,[7] Finally, in order to ensure efficient and constant supervision, the Directive mandates the creation of a national independent regulator, which must monitor the overall activities of the network companies and control network tariffs.

In 2005 the last Directive was published that directly affects the electricity sector.[8] This Directive establishes a framework where Member States must draw up transparent, stable and non-discriminatory security policies for electricity supply.[9]

Reality, however, shows that the theoretical principles are not so firmly consolidated and competition is not a fact in many Member States. In such a mixture of European regulatory models there are different market concepts and, as we said before, even the absence of competitive markets or the reduction to their minimum expression (Rivero, 2006).

Therefore, the present scenario of the European electricity sector is still far from being the competitive market expected and which was the fundamental aim of the liberalization process begun in 1990s. With some exceptions (the United Kingdom and the Nordic countries) what has happened in large countries such as France, Italy or Germany gives a good indication of the electricity sector liberalization process. These countries have kept their markets virtually closed, without effective competition, relying on public or private organized monopolies, either at a national or a regional level. Thus, the aspects related to energy should cease to be an exclusive object of sector policy makers and become the focus of attention by the European Union Competition Defence Authorities. According to the report on Competition published by the European Union on the liberalization process in the electricity sector,[10] such a process has not been successfully carried out. This report indicates that the power of monopolies has not been affected in all this time[11] and the non-collusion agreements among the main European energy groups have emerged as one of the most important confirmations, which eventually could pose a clear danger for the consumer (Ruiz, 2003).

On similar lines, the different Reports of the Commission (2004, 2005, and 2006) state their opinion on the start of an internal gas and electricity market. These reports show that, even though there are alternatives both in electricity and gas, the number of consumers who switched supplier up until 2005 in most Member States did not reach 50 per cent and, besides, many of those clients were dissatisfied with the range of products offered to them (European Commission, 2006). Regarding the progress of internal gas and electricity market implementation, it is still easily verifiable that, although at the beginning liberalization entailed an improvement of the efficiency in the energy supply and distribution and an important saving for the consumer, the later increase in prices has brought about no positive repercussion on the consumer.

5. Analysis of the liberalization process in the electricity sector

With the purpose of inquiring into the level of fulfilment of the Community Directives on liberalization in the European electricity sector, we have analysed different indicators which allow us to verify up to what extent this process has resulted, as intended, in a higher competition level of the sector.

5.1 Market design

The first question to ask in order to tackle the design of the European electricity market is what part is open to competition? There may be a wholesale competitive market where companies compete for gaining power over other generators; but the main objective of the European Union policy is to encourage more competition in retail markets.

An indicator of the development of competition in the retail electricity market is the opportunity for consumers to choose supplier. A direct method of measuring this is through the percentage of end consumers who have switched electricity supplier or renegotiated their contracts since the opening of the market. Table 5.2 shows the behaviour of consumers in this respect in the different Member States.[12]

In 2006 in most of the EU-15 countries and Norway all consumers can freely choose their supplier.[13] The exceptions are France, Italy, Belgium, Greece and Luxemburg, although freedom of choice in these countries also affects a significant percentage of consumers; but small businesses and household consumers have no opportunity to choose electricity

Table 5.2 Indicators relevant to market design, 2005

Country	Market opening[a]	Type of unbundling TSO	Type of unbundling DSO	Market model	Switched by consumer group[b] Large and very large industrial users	Switched by consumer group[b] Small to medium industrial users and businesses	Switched by consumer group[b] Very small businesses and households
Austria	100%	leg.	leg.	bilateral	29%	29%	4%
Belgium[c]	90%	leg.	leg.	bilateral	20%	10%	
Denmark	100%	leg.	leg.	hybrid	>50%	15%	30%
Finland	100%	own.	acc	hybrid	>50%	82%	0%
France	70%	leg.	man.	bilateral	15%		5%
Germany	100%	leg.	acc.	bilateral	41%	7%	0%
Greece	62%	leg.	none	bilateral	2%	0%	9%
Ireland	100% (56%)	leg.	man.	bilateral	56%	15%	n.a.
Italy	79% (56%)	own.	leg.	bilateral	60%	3%	0%
Luxemburg	84% (79%)	man.	man.	bilateral	25%	n.a.	11%
Netherlands	100% (57%)	own.	leg.	bilateral	n.a.	16%	19%
Portugal	100%	own.	acc.	bilateral	25%	22%	
Spain	100%	own.	leg.	pool			

Sweden	100%	own.	leg.	hybrid	>50%	n.a.	29%
UK	100%	own.	leg	bilateral	>50%	>50%	48%
Norway	100%	own.	leg./acc.	hybrid	>50%	>50%	44%
Estonia	12% (10%)	leg.	leg.	bilateral	15%	0%	
Latvia	76%	hh	hh	bilateral		0%	
Lithuania	74%	leg.	leg.	bilateral	15%		
Poland	80% (52%)	leg.	acc.	bilateral	19%		0%
Czech Republic	74% (47%)	leg.	acc.	bilateral	5%	1%	0%
Slovakia	77% (66%)	leg.	man.	bilateral	n.a.		0%
Hungary	67%	leg.	acc.	bilateral		32%	
Slovenia	75%	leg.	acc.	bilateral	8%	2%	0%
Cyprus	35%	man.	none	bilateral		0%	0%
Malta	0%	n.a	Single buyer	n.a.		0%	

Notes: TSO: Transmission system operator; DSO: Distribution system operator; own: ownership; leg: legal; man: management; acc: accounting.

[a] The data between brackets refer to 2004. In the rest there have been no variations.

[b] Volume of electricity consumption having switched by group cumulative since market opening.

[b] 1. Including switching between affiliates of the same group of companies or change from a standard regulated contract to an individually negotiated contract.

[b] 2. Italy and Spain include all consumers having left regulated tariffs.

[c] The data for Belgium refers to the Flemish region only.

Source: European Commission (2006).

supplier.[14] However, from the point of view of the European Commission total opening is an essential condition, although not sufficient, to ensure competition in the electricity sector. Despite having the opportunity of choosing supplier, the percentage of users who have made effective use of the freedom of choice is still low in most countries and the selection of a new supplier coming from another Member State is an exception. As was expected, the switch of supplier has been more common in the group of big industrial consumers and it is still rare in the group of small businesses and household users, for whom the opportunity to choose supplier is more recent; hence it is too soon to assess that behaviour. Considering the percentage of large clients who have switched supplier since the beginning of the opening of electricity market to 2006, we can differentiate a group of more advanced countries in the process of market liberalization, where more than half the qualified consumers have switched suppliers. In this new group we find the Nordic countries and the United Kingdom, highlighting the presence of Ireland and Italy. At the other end we have countries such as Greece, where only 2 per cent of consumers have switched supplier.

With regard to the Member States which have recently joined the European Union, their delay in the opening of the market leaves them in the group of countries with a lower percentage of consumers who have decided to switch supplier.[15] Nevertheless, Hungary deserves special attention since 32 per cent of its large and medium consumers have exerted their right to change supplier, surpassing countries where the opening of their markets to these collectives had already taken place several years before, such as Spain, Luxemburg and France.

There are several reasons that contribute to this slow development of the electricity market. One factor hindering the switch of supplier is that on many occasions there are no competitive offers or the offers are too similar to be a true option. Another factor is the existence of administrative barriers which prevent consumers from having significant information about prices and services rendered and sometimes even the contractual regime is not transparent enough. Finally, a clearly determining factor of the end-users' behaviour and which discourages change is the existence of regulated prices and particularly when they are kept at a lower level than that fixed by the market; these regulated prices sometimes do not reflect the real costs of the service rendered, which in turn hinders the admission of new supply companies to the market.

A competitive electricity market requires the good functioning of the wholesale market. Three categories have been established with reference to electricity market institutions, depending on whether they are based on a pool (Power Exchange), bilateral contracts (over the counter, OTC), and a third model, called hybrid, which is the mix of the two others. Power Exchange is an 'organized' and standardized marketplace on which all supplies and demands simultaneously meet on the basis of detailed rules fixed in advance. In bilateral markets a buyer meets one supplier at a time in direct negotiations. OTC markets are predominant in most of the countries such as the United Kingdom, France and Italy. The market based on a pool plays an important role in countries such as Ireland, Spain and Lithuania.[16]

Depending on the delivery period, bulk electricity can be traded on spot or forward markets.[17] Spot markets are day-ahead markets (single hours or blocks of hours) and intra-day markets on which electricity is traded one day before physical delivery takes place. On forward markets, power is traded for delivery further ahead in time (weekly, monthly, quarterly and yearly products). Spot prices on power exchanges are usually set in auctions, separately for 24 individual hours. Each market participant hands in price–quantity pairs for its selling and purchasing plans, from which the exchange derives aggregate supply and demand curves. The market price and the corresponding clearing quantity are then set as a result of the matching process. In comparison, on OTC markets spot transactions are carried out in continuous trading.

The European Commission published the Final Report of the Sector Inquiry in January 2007. This report shows, among other things, electricity volumes traded on spot and forward markets relative to electricity consumption in relevant European markets, so much on power exchange as OTC markets.

Table 5.3 shows that there are important differences between countries with regard to the size of spot markets. Relating to power exchanges, countries can be divided into two broad groups. In the first group members of power exchanges have some kind of obligation or incentive to trade via the exchange (OMEL, GME, Nord Pool). In the second group exchange participants have no such incentives or obligations (EEX, APX, Powernext, EXAA, and the UKPX). In this group EEX and APX saw significantly higher spot volumes traded than the rest power exchange. The figures also shows that traded spot volumes on power exchanges are larger than OTC markets in most of the countries.

Because volatility of spot prices, sellers and buyers engage in forward contracts. They prefer price certainty to unknown spot prices in the future. Table 5.4 offers total traded volumes in forward contracts. In Spain forward trading is insignificant; on the contrary, the Dutch and German OTC forward markets traded the highest volumes.

Table 5.3 Spot traded volumes as a percentage of national electricity consumption (June 2004–May 2005)

Countries	Power Exchanges	OTC
Spain (OMEL)	84.0	negligible
Italy (GME)	43.7	n.a.
Nordic Region (Nord Pool)	42.8	n.a.
Germany (EEX)	13.2	5.4
The Netherlands (APX)	11.9	5.9
Belgium	No power exchange	0.1
France (Powernext)	3.4	1.5
Austria (EXAA)	3.0	n.a.
United Kingdom (UKPX)	2.1	8.6

Source: European Commission, 2006a, pp. 126.

Table 5.4 Traded volumes in futures/forward contracts as a percentage of national electricity consumption (June 2004–May 2005)

Countries	Power Exchanges	OTC	Power Exchange + OTC
Spain (OMEL)	No exchange trading	negligible	n.a.
Italy (GME)	No exchange trading	n.a.	n.a.
Nordic Region (Nord Pool)[a]	196	327	523
Germany (EEX)	74	565	639
The Netherlands (Endex)	39	509	548
Belgium	No exchange trading	22	22
France (Powernext)	6	79	85
Austria (EXAA)	No exchange trading	n.a	n.a.
United Kingdom (UKPX)	0	146	146

Note: [a] This figure only includes bilateral contracts cleared by Nord Pool in 2005.
Source: European Commission, 2006a, pp. 127.

5.2 Transmission network

The second Directive on liberalization of electricity sector (2003), establishes the necessary vertical unbundling between regulated activities (transmission and distribution) and activities open to competence (generation and supply). The Commission fixes four unbundling levels: ownership separation, legal separation, management separation and accounting separation. Ownership and legal separation are considered to be the best conditions in order to achieve effective competition. Thus, tariffs reflecting real cost and the elimination of any cross-border subsidies can be guaranteed. In addition, one of the conditions established is the network operator independence, which allows equitable third-party access to the network. A regulated third-party access system based on public tariffs will lead to a more competitive market given the transparency of prices, ensuring the non-discrimination of the utilities which gain admittance to the network.[18] In general, but with some exceptions, this is the system applied by European countries.

In most countries there is one transmission system operator and several operators for the distribution system.[19] Table 5.2 above showed the kind of activity unbundling that has been carried out in different countries. As it can be seen, several countries meet the optimum condition to safeguard a more competitive market. As for the transport network, the countries having an ownership separation from the system operator are: the Nordic countries, the United Kingdom, Portugal, Spain, Italy and the Netherlands. Except for Luxemburg, whose separation is only one of management, the rest of the countries have legal separation. On the other hand, there is no single case of ownership separation in the distribution network, where the most common separation is the legal one.

5.3 Business concentration

Liberalization of the electricity market in many EU countries began from a monopoly or oligopoly structure. The introduction of competition was intended to put an end to this situation, though it has not been achieved in most markets. On the contrary, concentration in domestic markets is still high and in some cases, since the opening of the market began, the sector has become more concentrated through mergers and takeovers both at a national and cross-border level, which may have limited competition.

An indicator that reflects the level of competition in national markets is the number of companies which develop their activity and their total market share. Table 5.5 gives information about the structural market in the potentially competitive activities: generation and supply. Generally speaking there is a high concentration level with a very low number of companies responsible for most of the electricity generation and supply to clients. In some countries, markets are controlled by one or two companies and there is usually not enough capacity for cross-border competition. Moreover, on the whole, countries show a high correlation between concentration level in generation and supply. This trend

Table 5.5 Indicators of structural market, 2005

	Wholesale Market		Retail Market	
	Number of companies with at least 5% share of production capacity	Share of largest 3 producers share over 5%	Companies with market companies	Market share of largest 3
Austria	5	54%	4	67%
Belgium	2	95%	2	90%
Denmark	10	40%	5	67%
Finland	10	40%	6	30%
France	1	96%	1	88%
Germany	5	72%	3	50%
Greece	1	97%	1	100%
Ireland	2	93%	4	88%
Italy	5	65%	6	35%
Luxemburg	1	88%	2	100%
Netherlands	4	69%	3	88%
Portugal	3	76%	3	99%
Spain	3	69%	5	85%
Sweden	10	40%	4	70%
UK	8	39%	6	60%
Norway	10	40%	4	44%
Estonia	1	95%	1	n.a.
Latvia	1	95%	1	99%
Lithuania	3	92%	1	100%
Poland	7	45%	3	32%
Czech Republic	1	76%	8	46%
Slovakia	1	86%	4	84%
Hungary	7	66%	7	56%
Slovenia	3	87%	6	71%
Cyprus	1	100%	1	100%
Malta	1	100%	1	100%

Source: Regulators' data. Taken from European Commission (2006a).

is probably caused by the strong vertical integration of both activities in most domestic markets.

In principle, if EU countries advance towards larger regional markets or the single European market, national concentrations may cease to be relevant, provided there is an adequate interconnection capacity. In the short run, though, evidence indicates that markets will continue to be primarily national (especially in the largest markets such as the United

Kingdom, France, Germany, Italy and Spain), where cross-border interconnection network is not strong enough to reduce domestic market power (Jamasb and Pollitt, 2005). An indicator of the level of market concentration is the number of enterprises generating at least 5 per cent of electricity. Table 5.5 shows the number of companies which contribute that quantity is not higher than 10 in any country; and that figure is only reached in the countries with a more advanced level in market liberalization, such as the Nordic countries. However, the number of countries where three or fewer companies are responsible for most electricity production and supply is especially high. Such is the situation of Belgium, France, Greece, Luxemburg and most of the European newcomers.

When the share of electricity production and supply of the three largest companies operating in a country are taken as indicators of the market concentration level, countries can be classified in three categories: moderate, intermediate and high concentration levels.[20]

1) *Moderate concentration level.* Electricity markets in Finland, Norway and Poland have a moderate concentration level in potentially competitive activities. Denmark, Sweden and the United Kingdom also show a low concentration in the generation market but an intermediate concentration in the retail market, with the three largest enterprises concentrating over half the electricity supply.
2) *Intermediate concentration level.* Markets in Austria, Germany, Italy, Spain, Portugal, the Netherlands and Hungary are in an intermediate position in terms of concentration in the wholesale market. They also have a similar concentration level in retail market except for Italy and the Czech Republic, where concentration level is moderate (35 per cent and 44 per cent, respectively) and, on the contrary, in Portugal concentration in this market is very high (99 per cent).
3) *High concentration level.* Belgium, France, Greece and the rest of the EU newcomers show a high concentration level. In these markets one or two enterprises are responsible for almost the whole electricity production and trade.

High concentration has allowed incumbents a dominant position in their domestic markets with situations in which they can only enjoy relevant information to negotiate in the market. This has made the access of new operators difficult, especially in the activity of electricity generation, while at the same time limiting competition. On the other hand, market control has obliged many Member States to carry out a regulatory intervention, keeping a strict control of electricity prices which end-users pay.

The European Commission admits these regulated tariffs may be justified by the need to protect consumers against price manipulation but it also admits that it hinders competition and the smooth running of domestic electricity markets. Moreover, it prevents prices from acting as signals of new capacities and on discouraging investment; they put security supply at risk (European Commission, 2006a).

Many countries are studying ways to face concentration without resorting to structural adjustment reducing their generation capacity (disinvestment or transmission). The European Commission proposes several measures to reduce the adverse effects of high concentration. A possible line of performance would be to demand more transparency from electricity generation companies about the current and foreseeable capacity as well as some guidelines for the methods used by dominant generation companies to offer energy. In the most serious cases of concentration problems, another possible measure would be the creation of supply virtual auctions (i.e. Electricity Virtual Plant Auctions Programmes).

5.4 Price evolution

On increasing domestic and cross-border competition, liberalization should contribute to improving energy supply and savings for customers. Effective regulation of transmission networks also permits cost reduction: suppliers can interact more efficiently with clients and offer a wider range of services and contract agreements. All this implies that in the medium and long term competition should represent a price reduction for end-users.

Data relative to electricity price evolution between 1995 and 2006, both for households and companies show however that these expectations about price reduction have not been fulfilled. Figures 5.3 and 5.4 reflect this, although at the beginning of the liberalization process of the sector in different countries prices were reduced, once the initial effect passed off they began to rise, partly offsetting the previous declines. One of the reasons for the increase in prices, especially since 2000, was an increase in the prices of different primary energy sources, above all, fossil fuels used in electricity generation.[21]

Table 5.6, which shows the electricity price evolution for end-users, corroborates the information stated earlier. Between the years 1995 and 2000, when the first steps towards European liberalization were taken, there was a decline in electricity prices both at household and industrial level. However, it is paradoxical to note that once the electricity domestic market was consolidated, prices kept on rising rather than decreasing in

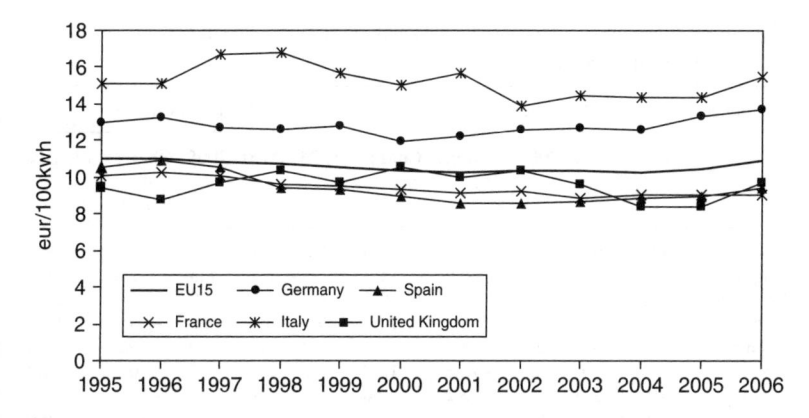

Figure 5.3 Electricity prices for households (all taxes excluded) (current prices in € per 100 kWh)
Source: Eurostat. Our estimation.

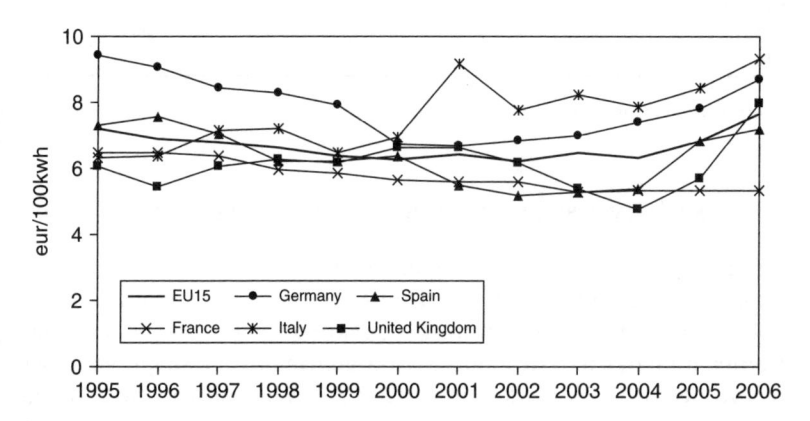

Figure 5.4 Electricity prices for industrial users (all taxes excluded) (current prices in € per 100 kWh)
Source: Eurostat. Our estimation.

both sectors, with the exception of the United Kingdom, France, Belgium and Austria. On the other hand, price divergences among countries are really wide, especially in the prices household consumers pay. These price differences have remained unchanged for the analysed period, and have even increased; consequently price convergence has not taken place.

180 Critical Essays on the Privatization Experience

Table 5.6 Electricity prices for households and industrial (all taxes excluded) (current prices in € per 100 kWh)

Countries	Household prices					Industrial prices				
	1995	2000	2005	Change (%) 95/00	Change (%) 00/05	1995	2000	2005	Change (%) 95/00	Change (%) 00/05
EU-15	11.0	10.3	10.9	−6.4	1.0	7.2	6.3	7.7	−13.3	9.1
Belgium	12.3	11.7	11.2	−4.9	−4.7	7.8	7.3	8.3	−5.4	−5.3
Denmark	6.1	7.2	10.0	18.1	29.1	4.3	5.0	7.2	16.4	28.2
Germany	13.0	11.9	13.7	−8.2	12.0	9.4	6.8	8.7	−28.5	15.6
Ireland	7.3	8.0	12.9	8.3	50.6	6.3	6.6	10.0	5.2	35.3
Greece	6.5	5.6	6.4	−12.8	12.9	5.7	5.7	6.7	0.7	13.0
Spain	10.6	9.0	9.4	−15.2	0.6	7.3	6.4	7.2	−13.0	7.9
France	10.1	9.3	9.1	−7.8	−2.5	6.5	5.7	5.3	−12.8	−6.0
Italy	15.1	15.0	15.5	−0.6	−4.0	6.3	6.9	9.3	9.3	21.6
Luxemburg	10.7	10.6	13.9	−1.0	22.0	7.7	7.1	8.5	−7.3	6.1
Netherlands	8.5	9.4	12.1	10.9	17.5	6.0	6.7	8.6	12.1	20.5
Austria	n.a.	9.5	8.9	n.a.	1.6	8.1	n.a.	6.5	n.a	n.a.
Portugal	12.6	11.9	13.4	−5.0	10.0	80	6.4	8.2	−19.5	10.9
Finland	7.0	6.5	8.1	−8.3	22.8	4.5	3.8	5.2	−16.0	39.8
Sweden	n.a.	6.4	8.8	n.a.	32.8	n.a.	3.8	5.9	n.a.	23.2
UK	9.5	10.6	9.7	11.6	−20.8	6.1	6.6	8.0	9.6	−14.2

Source: Eurostat. Our estimation.

6. Electricity restructuring in Europe: large countries

Electricity privatization and deregulation in the European countries have happened in parallel to the restructuring of the industry, differentiating between generation, transmission, distribution and supply activities. Competition has been introduced in the generation and supply extremes of the electricity chain, keeping transmission and distribution as regulated activities. However, despite institutional efforts to divide the electricity sector and reduce concentration, private companies have managed to increase their size through mergers. Thus, in many countries the supply business is completely integrated with that of generation, leading to a vertically integrated structure.

Privatization and restructuring processes happening together with liberalization of the electricity sector in European countries have permitted their classification in three groups: first, those countries which have opted for the market without reserves, with relatively small concentrated structures, such as the United Kingdom and the Nordic countries; on the contrary, we could find those countries with a slow regulatory

Table 5.7 Largest European electricity companies, 2006

Company	Stock market capitalization (€Million)
EDF (F)	111 881
E.ON (G)	69 262
Suez (F)	50 841
Enel (I)	48 583
RWE (G)	44 070
Endesa (ES)	40 920
Gaz de France (F)	34 081
Electrabel (B)	32 098
Iberdrola (ES)	31 914
Céntrica (UK)	20 523
Fortum (N)	20 179
Scottish & Southern (UK)	19 670
Scottish Power (UK)	17 496

Notes: F = France, G = Germany, I = Italy, ES = Spain, B = Belgium, UK = United Kingdom, N = Norway,
Source: Bloomberg. From *Expansión*, 24 March 2007.

development, which have decided to protect their companies, favouring the big 'national champions', such as France.[22] The rest of the countries are in an intermediate situation, with different evolution paces.

Some governments have done their utmost to protect their companies by promoting their concentration with the aim of making acquisitions by foreign companies less accessible. At the same time, those giants have acquired companies from other countries becoming 'European champions'. The existence of free circulation of capital in the European Union favours this situation. The result is that a few quasi-monopolies, some of them public such as EDF, ENEL or EDP or private such as E.ON or RWE, supported by their governments have acquired companies in other EU countries. As a consequence, Europe has fewer and fewer electricity companies, which weakens competition. Experience shows, on the other hand, that the least liberalized EU countries are the ones that receive less investment from overseas, and on the contrary, they are the ones with more investment initiatives in other EU countries (such as France with EDF). Table 5.7 shows the largest electricity companies in the EU and Table 5.8 their presence in other European Community countries. Later, we will focus the study on a deep analysis of concentration and privatization processes of companies which have

Table 5.8 Presence of the largest companies in other Member States

	Largest	Others (main ones)
Austria	VERBUND	RWE, E.ON, EDF
Belgium	E-BEL	EDF, ESSENT, NUON
Denmark	ELSAM	E2, VF, E.ON
Finland	FORTUM	VF, E.ON
France	EDF	E-BEL, ENDESA
Germany	RWE	E.ON, VF, EDF
Greece	PPC	
Ireland	ESB	NIE (Viridian)
Italy	ENEL	E-BEL, ENDESA, EDISON, VERBUND
Netherlands	E-BEL	ESSENT, NUON, E.ON
Portugal	EDP	ENDESA
Spain	ENDESA	IBERDROLA, EDP, ENEL, U. FENOSA
Sweden	VF	E.ON, FORTUM
UK		EDF, E.ON, RWE, IBERDROLA, CENTRICA
Poland	BOT	PKE, PAK, E-BEL, EDF
Czech. Rep.	CEZ	RWE, E.ON
Slovakia	ENEL	TEKO, RWE, EDF, E.ON
Hungary	MVM	EDF, E.ON, RWE

Notes: E-BEL : Electrabel; VF: Vattenfall. Iberdrola has a presence in the UK after its merger with Scottish Power in 2007.
Source: DG TREN. Quoted from COM (2004) 863 final.

taken place at the same time as electricity liberalization in the largest European markets, that is to say, the UK, Spain, Germany, Italy and France.

6.1 Electricity market in the United Kingdom

Among developed economies the British electricity sector was one of the first to be privatized at a time when numerous state-owned companies were being sold. Because of its size and importance, electricity privatization was the biggest and most radical of the ones the UK tackled after the publication of the Electricity Act in 1989 (Beder, 2005, p. 419). Generation and supply sectors were liberalized, whereas transmission and distribution were maintained as natural monopolies. For the regulated segments a new authority was established, the Office of Energy Regulations (Ofgem).[23] Liberalization was completed in 1998 when all consumers were given the choice of supplier.[24]

In 1990 the wholesale market was organized around a *pool,* in which daily electricity generators introduced offers in order to meet the demand for each of the 48 half-hourly blocks of the following day. The pool was completed with a bilateral market (over the counter) where made-to-measure contracts were negotiated. This model began working efficiently but over time it received much criticism, which obliged the system regulator (Ofgem) to change its design radically. As a consequence, the pool disappeared in 2001 and the New Electricity Trading Arrangements (NETA) was set up, later called British Electricity Trading and Transmission Arrangements (BETA) when it joined the Scottish system. In this new design almost the whole buying and selling is organized around bilateral contracts between generation and demand.[25] In parallel a new daily market emerged (day-ahead market),[26] the APX Power UK, where half-hourly blocks are continuously negotiated. Likewise, the market operator, ELEXON, manages the balancing market, where offers are introduced to offset the imbalance.[27]

Before privatization, in England and Wales, the state-owned Central Electricity Generating Board (CEGB) was the owner of the electricity generation power stations and transport networks. The first step towards privatization was the restructuring of CEGB in 1990. The 12 regional distribution management units were transformed in new Regional Electricity Companies (RECs), responsible for electricity distribution and supply to consumers in specific regions.[28] The national transmission network came under the control of the National Grid Company (NGC), whose ownership was initially transferred to the RECs and, later in 1995, was quoted on the Stock Exchange. On the other hand, generation assets were distributed among three new enterprises: National Power, PowerGen and Nuclear Electric (which was later divided in two: British Energy and British Nuclear Fuel). The two first companies were quoted on the Stock Exchange, whereas Nuclear Electric remained state-owned until its privatization in 1996. Although, in 2002, British government again purchased the company in order to avoid its bankruptcy. In Scotland, public enterprise was divided into two completely integrated companies, which later became privatized: Scottish Power and Scottish Hydro-Electric.

Shortly after privatization, the electricity sector structure began to change drastically. After the price manipulation in 1994,[29] the regulator urged the two largest generating companies (National Power and PowerGen) to initiate a disinvestment process by selling part of their plants to reduce their market share.[30] In return for selling part of their generation capacity these companies were allowed to acquire supply businesses from the RECs. This made the system become more integral: generation companies started to have a stake in the supply business at a significant level and the RECs could be integrated in generation. In the last ten years there has been an evolution from a monopoly structure to a break-up in smaller sized companies, with about 40 generation companies (Pérez Arriaga et al., 2006).

Nevertheless, efforts to divide industry with the aim of promoting competition have been offset by a surge of mergers and takeovers. The government maintained a 'Golden Share'[31] in many of the privatized companies. But when this governmental control of the RECs concluded in 1995, the aforementioned surge of mergers and takeovers took place. Moreover, British companies were taken over by foreign companies.[32] Initially, most of the purchasers were from the United States, but later they were replaced by the largest European companies. At present EDF (France) or E.ON and RWE (Germany) are owners of the largest electricity generating British companies.[33]

In 2002 the owner of the transport network, National Grid, merged with Lattice (formerly part of British Gas), creating National Grid Transco, a huge electricity and gas transmission company. There have also been attempts to integrate with other energy sectors with mergers between privatized electricity and water companies that operate in the same areas.[34]

The British electricity sector is presently dominated by six companies with important businesses in generation and supply, namely Npower, successor of National Power, owned by RWE; PowerGen, owned by E.ON; EDF; the two Scottish companies (Scottish Power[35] and Scottish & Southern Energy, the latter successor to Scottish Hydro), and a trading branch of the private gas company, Centrica, which operates in the United Kingdom as British Gas. These six companies control 53 per cent of the generation capacity in the UK.[36] From the rest, 20 per cent is owned by the two nuclear power stations, British Energy and British Nuclear Fuel (BNFL) (Thomas, 2005). These six largest companies also control the retail market, whose market shares range between 10 per cent and 25 per cent. (Pérez Arriaga et al., 2006).

6.2 Electricity market in Spain

Liberalization in Spanish electricity market began in 1997 when the Electricity Sector Act 54/97 came into force. It transferred Directive 96/92 CE on common norms for internal electricity market into Spanish legislation. This law represents a complete transformation of the electricity sector, limiting governmental action. Transmission and distribution activities are still regulated, given their monopolistic natural trend, whereas supply and generation activities are liberalized.

In order to ensure the functioning of the system there are two operators: the 'market operator' (the Spanish Electricity Market Operator, OMEL),[37] which assumes the economic management of the production market and the 'system operator' (Red Eléctrica Española, REE),[38] which is in charge of the technical management of the electricity system. Besides, the Energy National Commission (CNE) acts as regulator.[39]

In January 2003 the whole market became liberalized, a process which entailed the opportunity for all consumers to choose electricity supplier. This meant a four-year advance with regard to the initial timescale fixed by the European Union, which marked 1 July 2007 as the deadline.

The main feature of the Spanish electricity sector is its big concentration, since only five companies control the business: Endesa, Iberdrola, Unión Fenosa, Hidrocantábrico and Gas Natural. The two largest companies, Endesa and Iberdrola,[40] concentrate about 65 per cent of the production.[41]

Two of the companies that develop their activity in the electricity sector were created from public capital and in the last decade they have gone through a privatization process. These are Endesa, the first Spanish electricity company, and Red Eléctrica Española (REE), owner of the high tension transport network and natural monopoly.

Endesa is the Spanish electricity leader company. It was set up in 1983 as a state-owned company and the government began its privatization in 1988 through consecutive quotations on the Stock Exchange Market, completing this process 10 years later.[42] At present, the Sociedad Estatal de Participaciones Industriales (SEPI) controls 3 per cent of the company capital, the rest of the capital being in the hands of private owners. Its importance in the sector has increased due to successive mergers and takeovers from other companies[43] It concentrates 38 per cent of the wholesale generation market and it is the largest private electricity multinational in Latin America and has a consolidated position in the electricity Mediterranean market, especially in Italy.

The other state-owned company to follow a similar privatization process is Red Eléctrica Española (REE), the only transmission system operator. It was created in 1985 mainly with public capital (51.4 per cent), and its ownership was shared by electricity companies Endesa, ENHER, Iberduero, Hidroeléctrica, FECSA, Unión Fenosa and Sevillana de Electricidad. Privatization of REE, with its quotation on the Stock Exchange, commenced in 1999 through a takeover bid by means of which SEPI (Sociedad Estatal de Participaciones Industriales) quoted 31.5 per cent of REE on the market.[44]

Since 2000 there have been several merger attempts, which have not materialized due to government or CNE opposition. In 2000, Unión Fenosa tried to gain control of Hidrocantábrico, but the government did not approve that merger. The same year there was an attempt to merge Endesa and Iberdrola, with an ambitious disinvestment plan, but the conditions imposed by the government were too tough and not accepted, consequently the merger attempt failed. In 2003, Gas Natural tried to take control of Iberdrola, but the operation was blocked by CNE veto.

In the last years, the electricity sector witnessed important integration movements with the arrival of foreign companies. In 2004, the Portuguese state owned company EDP (Electricidade de Portugal) became the strongest shareholder of Hidrocantábrico.[45] Moreover, EDF is indirect owner of the company through the German subsidiary EnBW. Likewise, the French company Suez is present in Gas Natural.

Unión Fenosa is also a potential target for takeover. In September 2005 ACS, the main Spanish construction company bought 22 per cent of the company, holding at present 40.5 per cent. Their aim was a future merger with other Spanish electricity companies, mainly with Iberdrola, in which the construction company ACS has its stake.[46] This possibility seems more distant since Iberdrola purchased, in April 2007, the British utility Scottish Power[47] and in July the American company Energy East. This merger has turned Iberdrola into the third largest electricity European company.

Finally, we would like to emphasize the successive attempts made to control the leader company in the sector, Endesa. This process began in September 2005 when Gas Natural launched a hostile bid for Endesa; later, E.ON did the same and finally the construction company Acciona and the electricity Italian company ENEL came into the conflict, interfering with the German consortium.[48] This odd process, featured by the mentioned companies, and several actions carried out directly by the government or through the electricity market regulator (CNE) and the Stock Exchange regulator (CNMV),[49] finished after E.ON withdrew its bid and the German company ENEL and Acciona reached an agreement to share out Endesa's assets. Consequently, the first electricity company will be divided up.[50]

The agreement reached favours the three companies: E.ON will hold part of Endesa's assets in Italy, France and Spain (including Viesgo, owned by ENEL) as well as several assets Endesa had in other European countries. The construction company Acciona (also the main Spanish supplier of wind energy) will obtain Endesa renewable energy business and ENEL will hold the rest of Endesa's assets, increasing its presence overseas. With the distribution of Endesa among three different owners, the attempt of the Spanish government to maintain the company under Spanish control failed.

At the beginning of March 2007 Spain and Portugal signed a regulatory coordination agreement to create the Mercado Ibérico de la Electricidad (MIBEL) and for the construction of new interconnections between both countries.[51]

6.3 Electricity market in Germany

The electricity market in Germany is the largest in the European Union in terms of electricity generation. Furthermore, the system is interconnected with neighbouring countries, although there are hardly any exports.

The European directive 1996 marked the beginning of the German electricity sector reform putting an end to a monopoly that had lasted over 100 years. In April 1998 Germany passed the Electricity Energy Act (EnWG, Energiewirtschaftsgesetz), which established the bases for the reform and which in some aspects was clearly different from the reforms carried out in the rest of the European markets. From the first moment, the market opened for 100 per cent of consumers (all consumers could freely choose supplier). Legal unbundling between electricity generation and transport was put forward; and in the same way they opted for negotiated access of third-parties to the network, rather than non-regulated access.

The key factor governing the development of the retail market at the start of the liberalization of the electricity sector was the negotiated (rather than regulated) network access tariffs. Any company willing to compete in the German market had to negotiate these access tariffs with the electricity companies already operating in the sector, unwilling to lose a single client, a situation that implied a strong barrier to access (Pérez Arriaga et al., 2006). According to EU sources (European Commission, 2005), 41 per cent of large consumers, 7 per cent of small businesses and 5 per cent of domestic consumers had switched supplier.

In 2000 two wholesale organized markets started to operate: the Frankfurt European Energy Exchange (EEX) and Leipzig Power Exchange (LPX). In 2002 both markets merged giving rise to the new European Energy Exchange (EEX), located in Leipzig. In its daily market trading of hour and block contracts are continuously negotiated.[52] Additionally electricity is traded over the counter (OTC) spot market but the largest share of the market is performed by forward contracts. Then in 2005

Table 5.9 Germany: Generation market shares (%)

Pre-merger	Post-merger	Pre-merger	Post-merger
VEBA	E.ON	18.77	28.74
VIAG		9.77	
RWE	RWE	28.94	37.27
VEW		8.33	
EVS	EnBW	8.60	8.60
Badenwerk			
HEW	Vattenfall	2.57	15.03
BEW AG		2.13	
VEAG		10.33	
Otros	Otros	10.35	10.35
TOTAL		100.00	100.00

Sources: Drasdo et al. (1998), Bergman et al. (1999), Brunekreeft (2003). Taken from Brunekreeft and Twelemann (2004, p. 5).

a new law established the creation of the 'regulator' figure in the electricity industry. This regulating body was called Federal Network Agency (Bundesnetzagentur), and one of its functions was the regulation of gas, telecommunication, post, and railway access.[53] Consequently negotiated access was changed to regulated access. At the same time, there was a two-year deadline to put an end to the regulated tariffs that the 16 federal states had designed for small consumers.

Most of the companies in Germany have been regional companies, which rule electricity generation and which own regional transmission networks. In 1990, with re-unification, in West Germany there were eight electricity generation companies whereas in East Germany there was only one company operating, VEAG, owned by western companies.[54] Later these companies were involved in merger operations.[55] In 2003 E.ON closed the purchase of the gas company Ruhrgas with the support of the German government, which wanted to ensure natural gas supply to Germany. Table 5.9 shows the pre-merger and post-merger situations.

Merger operations have allowed electricity sector activities to be currently controlled by four companies: E.ON, RWE, EnBW and Vattenfall. The largest companies are E.ON and RWE, approximately the same size, which control 60 per cent of electricity generation and approximately 60 per cent of end-users' supply (Thomas, 2005). The four companies control over 75 per cent of the generation market share, which is highly concentrated and is similar to a vertically integrated regional monopoly structure: RWE in the northeast, EnBW in the southeast, E.ON in a north–south stretch in the middle of the country and Vattenfall in the new Länd Hamburgo-Berlin. These companies, owners of the transmission network in Germany, are the operators of such systems in their respective geographical areas. These companies are additionally characterized by high vertical integration, with activities not only in generation but in distribution and supply also. Approximately 10 per cent of electricity is generated by regional enterprises, which in turn are integrated, totally or partially, in large electricity companies. About 14 per cent of electricity is generated by self-producers (OECD, 2004).

The distribution sector is more complex. There are about 1200 regional and local distribution companies. Many of these companies are totally or partially owned by large companies. In the retail market there are also a large number of enterprises, among them many foreign companies such as Vattenfall, EDF, Electrabel or ENEL. RWE owns a 5 per cent quota, although taking into account the local companies it controls its quota may reach 30 per cent or even more. E.ON supplies electricity only to large industrial consumers but it also has some stake in many regional supply companies. Its quota in the national retail market is estimated to be about 32 per cent, without considering the regional companies it owns (Thomson, 2005).

6.4 Electricity market in Italy

One of the features that distinguishes the reform of the Italian electricity system is the will to maintain the power of the formerly monopolistic company ENEL (Ente Nazionale per L'Energia Elettrica), currently semi-private (the state owns 31.5 per cent of the capital). Additionally the measures have been implemented slowly and until 2004 there was no (more or less) complete regulation framework.[56]

The origin of the Italian electricity system reform dates from 1999, when the decree law, Decreto Bersani,[57] was published. It applied the European Directive 96/92 and a new market, quite similar to the Spanish one, was created (Gestore del Mercato Elettrico, GME).[58] Likewise they established an operator of the system (Gestore della Rete di Trasmissione Nazionale, GTRE) whereas ENEL withheld the ownership of the net through TERNA.[59] The system regulator was Autoritta per l'Energia Elettrica e il Gas. A timetable for liberalization of the retail market was set up and in 2000 this market was opened for large consumers, initially 30 per cent, who could choose supplier and from July 2004 only domestic consumers remained ineligible. They had to wait until July 2007 when the opening of the electricity market was complete.

ENEL privatization began with the sector reform in 1999. Then the state-owned company ENEL generated about 80 per cent of the country total electricity. The rest came from small local self-generator companies which sold their production to ENEL at regulated prices. The decree-law established that the maximum electricity generation quota of a company could not exceed 50 per cent, thus requiring ENEL to get rid of 15 of its 55 GW by 2002. ENEL did so through the sale of three companies of its own.[60] By 2003 ENEL had privatized 34.5 per cent of its company. Subsequently, it has continued the privatization process in different steps: in 2004 it sold 16.4 per cent and in 2005 11 per cent (Thomas, 2005). At present, the Italian government owns 31.5 per cent of the company capital. The company has also expanded abroad: in 2001 it acquired the Spanish company Electra de Viesgo, owned by Endesa. As previously noted, in March 2007, after the consecutive attempts made by utilities to gain control of the first Spanish electricity company (Endesa), ENEL bought 24.9 per cent of the company, hindering the bid launched by the German E.ON. This process finalized with the agreement reached by E.ON, ENEL and Acciona. ENEL was the only European company that did not operate or own nuclear reactors. This situation changed, however, in 2005, when it bought 66 per cent of the company Slovenske Elektrarne (SE) from Bratislava, which gave it access to nuclear generation capacity.

The sector is currently undergoing a period of transition. It seems that out of nine electricity generation companies only three or four will consolidate: ENEL, EDF/Edison and E.ON, which will enter the country once the agreement with ENEL and Acciona to buy Endesa Italia assets has been made effective. Local companies will probably play a relevant part as partners of the new groups.

6.5 Electricity market in France

The French electricity system is, in size, the second in the European Union, after Germany. 78 per cent of electricity comes from nuclear power stations, the highest percentage in Europe. The nuclear programme has contributed quite decisively to the fact that the French electricity system owns a strong over-capacity, which has helped it become the biggest European electricity exporter. Its privileged geographical location, with solid interconnections with its neighbouring countries, has favoured its export role.

France was the last European country (EU-15) to liberalize its electricity sector, giving strong support to the protection of the 'national champion' (French company that dominates the market). Its liberalization process began in 2000 when a new Law was introduced, reversing the European Directive of 1996.[61] This regulation established the unbundling of activities (generation, transmission, distribution and supply), the creation of a system regulator Comission de Régulation de L'Energie (CRE), the creation of the system operator RTE (Réseau de Transport d'Electricité), owned by EDF,[62] who were responsible for ensuring supply and access to the network and the gradual opening of the retail market until its total opening in July 2007.[63]

At the end of 2001 the wholesale market, Powernext, was created. It was initially made up of a day-ahead market where physical offers and demands are met in 4-hour blocks.[64] Subsequently an instalment market was developed, where one month-ahead, two month- ahead, or year-ahead contracts may be negotiated.

The evolution towards a competitive market has been slow and the generation market structure has not undergone significant changes. The state-owned company EDF (Electricité de France) is the largest European electricity company and one of the largest electricity producers in the world.[65] This company maintains a predominant position in the French market with over 90 per cent production share. In November 2005 the French government began EDF privatization. Its intention was to privatize 15 per cent of its capital, announcing it would not privatize more than 30 per cent in any case, at least, in the medium term.[66] In January 2006, its shareholding structure was divided in the following way: 87.3 per cent belonged to the state, 1.9 per cent to the company staff and 10.8 per cent was free float.

The most relevant local competitors to EDF are SNET (Sócieté Nationale d'Électricité et de Thermique), the second electricity production company controlled by Endesa, CNR (Compagnie National du Rhône) and SHEM (Société Hydro Électique du Midi), controlled by the Belgium company Electrabel.[67] EDF is also the dominant distribution company. In distribution there are a large number of local independent companies, about 170, but generally they are small and distribute electricity to 5 per cent of consumers. The most important companies are in Strasbourg, Metz and Grenoble.

In retail sale or electricity supply EDF also has a prevailing position. Other relevant traders are Poweo (linked to the Austrian utility Verbund) and Direct Energia (whose suppliers are Swiss EGL and the French branch of the Swedish company Vattenfall). The retail market development has been rather limited. In the first years, when only big consumers could opt for giving up the tariff, their arrival at a liberalized market was noticeable, encouraged by the significantly lower market price than the regulated tariff. Later on, though, the progressive rise of market price hampered this trend. By the end of 2005 less than 10 per cent of the consumers with opportunity to choose had decided to abandon the tariff. A third decided to contract their consumption from a trading company other than EDF.

EDF has an important presence abroad as owner of companies from other countries. In Europe it is present mainly in Germany, Italy and the United Kingdom. It owns 45 per cent of the capital of the German EnBW. In 2005, in Italy together with the local AEM Milan, it gained control of Edison, the second largest Italian electricity company and the third in gas generation. It has also been present in the United Kingdom since 1998, through its subsidiary EDF Energy, previously known as London Electricity and through its trading subsidiary EDF Trading. In recent years there have been some merger attempts between EDF and GDF (Gaz de France), but they have not been successful. GDF is the dominant company in gas sector. It is also a state-owned company but underwent a partial privatization in 2005. Among its present plans they are considering a merger with the French service group Suez.[68]

7. Conclusions

The analysis carried out on the main national electricity markets, as well as the level of fulfilment of the Community Directives on liberalization of the European electricity sector provides a picture of the present situation and the main obstacles to the single European market in the sector.

The first conclusion is that liberalization in European markets has not led, at least at the moment, to tougher competition. Experience shows that if a market has to be successful it is not enough with just some liberalizing norms; it is essential for the industry and market structure to be pro-competitive. Then the concentration problem is currently the most important obstacle for the development of more effective competition. This concentration, both vertical and horizontal, has been increased by the merger and takeover processes adopted by the sector largest companies, thus showing that Member States have not wanted or managed to tackle the matter of the market structure efficiently enough.

Europe started from a rather uncompetitive oligopolistic structure, accentuated with liberalization, since the performance of companies, both private and state-owned has not brought about more competition, but more concentration. Some countries have never dismantled the existing monopolies in this sector (France, with EDF; Italy, with

ENEL; Portugal, with EDP; or Belgium, with Suez-Electrabel), whereas in other countries such as Germany or Spain, after liberalization, duopolies or oligopolies have been established through concentration operations. In Europe, except for the United Kingdom and the Nordic countries, a 'national champions' policy has emerged: some companies with a predominant position in each country.

The failure of full integration of domestic energy supply in a larger European market is another urgent problem the sector must face. In this context the need to improve the norms related to electricity cross-border trade should be emphasized in order to make sure the existing infrastructures are used to the utmost, thus increasing competition. It is also necessary to promote investments in new interconnectors between national electricity systems but progress in this sense is still slow.

With regard to the unbundling of network operators and the introduction of a regulated access to third parties some progress has been achieved but some aspects are still unsatisfactory. In this respect, to make a market work correctly, the transmission and distribution network regulating authorities must be independent so that they may guarantee an equitable access of new suppliers to the network.

The existence of regulated tariffs for end-users is another hindrance worth mentioning. These tariffs, which are generally lower than market prices, may cause great harm to the creation process of a competitive market and reduce investment in it.

Regarding prices paid by final consumers, while at the beginning of the liberalization process prices decreased, once the initial effect passed off they started to rise partly offsetting previous reductions. Price divergences among countries are rather large, especially, in the absolute prices paid by households. These differences have been maintained in the analysed period and have even increased; consequently the desired price convergence has not taken place.

Finally, the different pace at which the countries are adopting the Community Directives for the attainment of the internal electricity market entails an obstacle for the market itself. The European Union does not seem to be willing to assume a relevant role in this matter. Its goal is focused on creating suitable conditions for the implementation of the single market, penalizing infringements and favouring cross-border flows, which the foreseeable increase of commercial electricity transactions will demand. In any case, it is not likely given the situation today that the Commission will be able to correct the regulatory unevenness existing among countries, which is essential for the creation of a true common energy policy.

Notes

1. When in 1973 General Pinochet overthrew the democratic Marxist government of Salvador Allende, he adopted some policies which were intended to reduce public deficit, privatize businesses and open the country's economy to the free market (Beder, 2005).
2. Since 1980s, the World Bank and the IMF have demanded public services privatization as a condition to grant loans and have prescribed policies so that the economies of these countries open up to multinational investment.
3. EIA (1996), p. 4.
4. In summer 2000 several supply problems arose in California, which made the State authorities intervene in the electricity sector in all its aspects, suspending the market-regulating norms which so far had been introduced as liberalization models. As Joskow (2006) indicates, this fact and the recurrent crisis in many other independent generation and supply companies, as well as the poor results achieved, especially for small consumers, cooled down the expectations existing about the electricity sector reform. Nevertheless, it has been proved that competition reached in wholesale markets contributed to greater transparency in price and tariff policies in the regulated market.
5. The European regulation aimed at promoting liberalization in the electricity sector of the EU countries originated in 1996 with the publication of Directive 96/92/CE on norms for the internal electricity market. This Directive established the general principles of the electricity sector liberalization. Particularly, it established the authorization procedures for the construction of generation capacity and provided for access conditions to the transmission and distribution network, which have to be objective, transparent and non-discriminatory. Member States had to enforce the necessary legal, statutory and administrative dispositions in order to fulfil what the Directive provided before 19 February 1999.
6. Directive 2003/54/CE, of the European Parliament and European Council of 26 June 2003, on common norms for the internal electricity market, a reason to abolish Directive 96/92/CE. It focuses upon the opening of the sector through stricter regulation of access conditions to the transmission and distribution network. Finally, the basis for a single European market is laid down. This Directive was to be applied by Member States by 1 July 2004.
7. With the aim of developing a single European market the EC Regulation n° 1228/2003, of the European Parliament and European Council of 26 June 2003 was published relating to access conditions into the network for cross-border electricity trade. The objective of this Regulation was to establish some equitable norms for cross-border trade, boosting competition in the internal electricity market. This implies the agreement on harmonized principles on cross-border transmission tariffs and on the allocation of the available interconnection capacity between national transport networks. This Regulation was to be applied by Member States by 1 July 2004.
8. Directive 2005/89/EC of the European Parliament and European Council of 18 January 2006, on the safeguard of security measures for electricity supply and infrastructure investment. This Directive was to be applied by Member States before 24 February 2008.

9. In 2006 the European Commission published The Green Book on Energy, a consultative document where the purposes of the European Commission in this matter are described.
10. Neelie Kroes, European Union Competition Commissioner requested a thorough investigation of the electricity and gas sectors in 2004. At that time she stated that all efforts made during the liberalization process had not provided benefits either to consumers or to European companies (Expansión, 3/01/07).
11. In fact, the behaviour of large energy companies such as E.ON and RWE (Germany), Gaz de France (France) and ENEL (Italy) have been denounced by the European Union Competition Tribunal (http://www.expansión.com, 10/01/07).
12. The Community Directive fixed 1 July 2004 as the deadline for total freedom of choice of electricity supplier for non-household clients, and 1 July 2007 for small consumers; nevertheless, many Member States opened the market for this collective several years before.
13. Total opening of markets has been progressively achieved: in 1998 Norway, Finland, Sweden, the United Kingdom and Germany; in 2001 Austria; in 2003 Denmark and Spain; in 2004 the Netherlands and Portugal; and Ireland in February 2005.
14. The opening of electricity market is complete in European countries in July 2007, except in Cyprus and Latvia, where the opening of markets is delayed until 2013.
15. In none of the new Member States was the opening of electricity sector complete in 2006; small consumers did not have the possibility of choosing supplier.
16. The most important European power exchanges are Nord Pool in Scandinavia, EEX in Germany, APX in Holland, Powernext in France, OMEL in Spain and GME in Italy.
17. Electricity for spot and forward delivery could be traded on both power exchanges and OTC markets. In most of the countries forward delivery on power exchange markets, which are considered spot markets, is not possible.
18. There are two different access systems to the network, the regulated and the negotiated systems. In the first, competent authorities fix tariffs, which are published and these are applied to all network users. In the second case, each network user negotiates the terms of its access with the system operator.
19. The Transmission System Operator (TSO) is responsible for the functioning of the high tension transmission network. The Distribution System Operators (DSO) are responsible for medium and low voltage distribution networks.
20. Moderate concentration level: less than 50% of electricity generation and supply. Intermediate concentration level: more than 50% and less than 85% of electricity generation and supply. High concentration level: more than 85% of electricity generation and supply.
21. Wholesale electricity prices are mainly influenced by plant availability, fuel prices, precipitation, wind speed, interconnector availability, temperature and CO2 certificate prices. End-users' prices also include network charges, balancing cost/capacity payment and retail supply margin.
22. France resisted pressure from the European Commission to privatize its electricity. However, in 2002, with the election of a centre-right government,

it finally agreed to open its electricity market to competition and partially privatize the state-owned company EDF (Electricité de France).
23. The Office of Electricity Market (Offer) was set up in 1989. In 2001 it merged with the Office of the Gas Market (Ofgas), establishing the Office of Gas and Electricity Markets (Ofgem).
24. After privatization, in 1990, 5000 large consumers – with needs higher than 1 MW – had the possibility of choosing supplier. This possibility was enlarged in 1994 to 50 000 consumers – whose needs were over 100 kW – and in 1998 all consumers were eligible, thus introducing competition (Thomas, 2002).
25. 90% of electricity is sold by generators through bilateral contracts (http/www.electricity.org.uk/about_ea/bic_pub).
26. The government expected no more than 10% of total electricity sales would take place in the daily spot market. In fact, the figure is much lower, and usually about 1% of the energy consumed is purchased in the spot market (Thomas, 2005).
27. In the last years, the present market design has been largely criticized. It has been suggested that bilateral markets lack transparency and have scarce liquidity. Moreover, there is an increasing concentration in the retail market and a strong vertical integration. This led all traders to raise their prices between 5% and 22% in the first semester of 2006, the highest increase ever seen (Pérez Arriaga et al., 2006).
28. Regional Electricity Companies (RECs): East Midlands Electricity; Eastern Electricity; London Electricity; Manweb; Midlands Electricity; Northern Electricity; Norweb; South Eastern Electricity (later Seeboard); South Wales Electricity; South Western Electricity; Southern Electricity; Yorkshire Electricity.
29. The system regulator (Ofgem) observed that since privatization had begun the costs had decreased owing to reduction in fuel prices, in capital costs for new plants and improvements in thermal efficiency. However, this reduction had not been reflected in the wholesale market prices. Actually, all generators had increased their prices and it was attributed to the market power exercised by generation companies, especially by National Power and PowerGen. (Beder, 2005).
30. First they were required to sell 6000 MW and, later, they were forced to discard other 4000 MW.
31. The 'Golden Share' allowed the British government to prevent anyone from owning more than 15% of the capital with the right to vote in a private company (Beder, 2005).
32. By 1997 only 5 out of the 12 RECs were still independent, and not owned by other companies (national or foreign). By 2000 only one REC was independent, Midlands Electricity, which was acquired by PowerGen in 2004.
33. By 1998 North American companies were owners of 60% of distribution and supply companies. However, after the California energy crisis in 2001, the number of US electricity companies taking part in mergers and acquisitions at an international level decreased and some North American companies left the UK. These companies were replaced by European companies, which became very active accounting for 77% of the international electricity trade in 2001. (Beder, 2005).
34. Both Norweb and South Wales Electricity merged with local water supply companies, whereas Scottish Power acquired Manweb (EIA, 1996).

35. In April 2007 the merger between Scottish Power and the Spanish Iberdrola was approved.
36. No single generator has more than 20% market share (Pérez Arriaga et al., 2006).
37. La Compañía Operadora del Mercado Español de la Electricidad (OMEL) is responsible for crossing supply and demand and fix base price in the wholesale market.
38. Red Eléctrica is the owner of the high tension network and has to guarantee that electricity reaches users in safe and reliable conditions.
39. Owing to this new wholesale market, Costs of Transition to Competition (CTCs) were established. The aim of CTCs was to compensate electricity generators for costs of past investments (made before 1998) because it was supposed that companies would suffer a decrease of income with the changing from a regulated system to a competitive one. They disappeared in June 2006.
40. Iberdrola is a private company created in 1991 through the merger of the electricity companies Iberduero and Hidroeléctrica Española.
41. These companies agreed to carry out virtual energy auctions jointly in April 2007, fulfilling the Royal Decree which fixes 2007 tariffs. This Decree obliges both companies to hold a series of auctions offering virtual power plant (VPP) capacity to transfer part of their production to other operators as a mechanism for increasing the electricity acquired through bilateral contracts and to boost competition.

    The Royal Decree 1634/2006 includes an additional disposition that required Endesa and Iberdrola to carry out the auctions, destined to any agent of the Spanish electricity market. Emissions would consist of options to purchase energy up to an established hourly power and with a price fixed beforehand.
42. 24.4% was sold in 1988, 8.7% in 1994, 25% in 1997 and 33% in 1998.
43. Before its privatization in 1996, it took control of Sevillana and FECSA. In 1999 it performed the merger of Empresa Hidroeléctrica de Ribagorzana (ENHER), Eléctricas Reunidas de Zaragoza (ERZ), Gas y Electricidad (GESA), Unión Eléctrica de Canarias (UNELCO), Electra de Viesgo (VIESGO) and Saltos del Nansa (NANSA). Nevertheless, Viesgo was sold in 2001 to the Italian company ENEL.
44. The new legal framework derived from the electricity sector Law of 1997 made REE open its capital. The law marks a maximum participation of 10% for all shareholders, so the four large electricity companies could not exceed 40% of the capital jointly. This maximum was not applied to SEPI, which had to maintain at least 25% until 2003. Subsequently, in 2003, the four large companies were obliged to reduce their participation in REE up to 3%, thus reducing their joint participation from 40% to 12% (REE, 2006). In 2005, with the publication of the Royal Decree Law 5/2005, new shareholding changes were announced. REE shareholders, except for SEPI, had to reduce their participation up to 1%. In September that year SEPI, which owned 28.5% of the company capital, sold 8.5% of the capital.
45. In this way, EDP gained control of Naturcorp, S.A., former state owned company of the Basque Country which was privatized in July 2003 and purchased by Hidrocantábrico. As a consequence, Naturcorp returned to public sector, in this case, Portuguese.

46. A characteristic of the electricity sector in Spain is the important presence of construction companies (ACS and Acciona, mainly). Their presence in the sector comes as a consequence of the huge capital sums accumulated during the construction boom, which lasted ten years, and who are at present are looking for shelter in the electricity sector (profitability and safety) because they foresee the decline of the construction sector.
47. Iberdrola's bid over the Scottish group has overcome all the regulatory obstacles, including the approval of shareholders, both in the Spanish and Scottish companies.
48. After the bid launched by Gas Natural, at the beginning of 2006, the German energy giant E.ON entered the contest by submitting a competition bid. In January 2007 there was a period of acceptance of both bids, but finally Gas Natural decided to withdraw. When it all seemed as if the operation would conclude with the takeover by E.ON, new actors came onboard. In an attempt to prevent E.ON from gaining control of Endesa, the construction company Acciona increased its stake, becoming the main shareholder with 21% of the shares. In February it was the Italian company ENEL, 31% of which was controlled by the government, the one which became shareholder of Endesa, with 24.9% of the shares, the maximum that can be reached without submitting a bid. Thus, it consolidated its position as the first shareholder followed by Acciona. The appearance of ENEL, added to that of Acciona, which controls 45% of Endesa's capital, made the success of E.ON bid difficult.
49. Although with different objectives, the Board of Directors of Endesa, on the one hand, and the government, on the other, triggered off the events. In the first case, the search for a company that could save Endesa from the Gas Natural bid was the reason why E.ON came on stage. On the contrary, the aim of the government to give shape to 'national champions' and to keep control of the first Spanish electricity company within the borders was the key factor for the appearance of Acciona in September. Likewise, the government is considered to be the party that boosted the subsequent entry of ENEL in the contest, and the promotion of a common cause with Acciona against E.ON, given the lack of national partners with enough financial capacity to share that struggle with the construction company against E.ON.
50. Considering the difficulty for E.ON in gaining control of Endesa, the German consortium decided to withdraw the bid, at the same time reaching an agreement with ENEL and Acciona at the beginning of April to buy Endesa's assets. This deal makes things easy for the electricity company and Acciona to take over Endesa, since they are not obliged to sign the six months' moratorium imposed by the Comisión Nacional del Mercado de Valores (CNMV) to submit its public bid.
51. Its aim is to have it operative before July 2008. These agreements include the merger of the two operators of the electricity and gas markets in Spain and Portugal in a single Iberian operator (OMI). This will be structured in two holdings: half to Spain (Omie) and the other half to Portugal (Omip), with 10% crossed shares and each owning 50% of the management market companies. Moreover, no individual shareholder can own more than 5% of these holdings and the companies of the sector will not be allowed to exceed a joint share of 40% of the capital.

With regard to the approval of new electricity interconnections between both countries it has been agreed that REE and its Portuguese counterpart REN (Rede Eléctrica Nacional) may exchange shares. On the other hand, the Spanish network operator had already announced an agreement to acquire 5% of the high tension Portuguese network from EDP. Published in *Expansión*, 'España y Portugal ponen a punto el Mercado Ibérico de Electricidad', 9 March 2007.

52. The electricity volume negotiated in daily market exceeds 10% (Pérez Arriaga *et al.*, 2006).
53. This body is based on the existing regulator in telecommunication, post and railway sectors.
54. VEAG (Vereinigte Energiewerke A.G.) was created at the beginning of 1991 through the merger of Vereinigte Kraftwerks A.G. Peitz and Verbundnetz Elektroenergie A.G. In 1996, RWE, VEBA and VIAG, each owned 25% of VEAG shares; the other remaining 25% was distributed among several western companies.
55. The main merger operations took place at the end of 1990s. E.ON was created in 1999 from the merger of VEBA (Preussenelektra) and VIAG (Bayernwerk), which were the second and third largest German companies respectively. At the same time, RWE, which was the most important company, merged with VEW, a small regional company. EnBW (Energy Baden-Wüterttemberg AG) was set up in 1997 from the merger between Badenwerk and EVS, and presently it is controlled by the French giant EDF. Finally, the Swedish company Vattenfall, through Vattenfall Europe, controls three companies, BEWAG, HEW and VEAG.
56. Legge 239/04: Legge di riordino del settore energetico, 23 August 2004.
57. Law Decree 79 of 21 March 1999, 'Decreto per la liberizzazione del mercato dell'energia elettrica'.
58. A daily, intra-day market and complementary services were created.
59. The new Decree Law demands that ENEL's participation in TERNA should not be higher than 20%.
60. In 2001 it sold ELETTROGEN, the second Italian electricity generation company. It was purchased by a consortium which included the local company, ASM Brescia, and the Spanish Endesa, which became owner of 51% of the company. ENDESA ITALIA was the result of this operation. In 2002 EUROGEN, currently EDINPOWER, was sold to a consortium dominated by the Italian Edison and the French EDF. Finally, in November 2003 it got rid of INTERPOWER by selling it on an equal basis to Energia Italia and to a consortium constituted by Electrabel and ACEA.
61. Loi n° 2000-108, relative á la modernisation et au develópement du service public d'electricité.
62. Although RTE is not legally separated from EDF, since 2002 it has worked independently in terms of management, accounts and finance. In August 2005 a Decree was passed establishing that RTE would be a completely separated company, owned 100% by EDF. The new company would be known as RTE EDF Transport (Décret n° 2005-1069 de 30 août 2005, approuvant les status de la société RTE EDF Transport).
63. In February 2000 the market opened to big consumers – those who used over 100GWh – representing 20% of the market. Subsequently, it has gradually

opened to other big consumers until in July 2004 it was liberalized for all companies, representing 68% of the market. By 1 July 2007 it achieved complete opening for all consumers.
64. The volume of operations of this market is still low. In 2003 scarcely 2% of French electricity demand was supplied in this market. In March 2005 daily average had increased up to 5%.
65. EDF was set up as a publicly-owned company in 1946. It is a completely integrated company with electricity generation, transmission, distribution and supply monopoly. EDF produces 22% of European Union electricity, mainly from nuclear energy. Its sources of electric energy are: nuclear, 74.5%; hydroelectric, 16.2%; thermal, 9.2%; wind and renewable, 0.1% (http://www.edf.com).
66. In April 2005 the French government announced its plans to sell 30% of its shares in EDF. However, in September it announced it would not sell more than 15% to the public, reserving the other 15% for the employees of the company (http://www.edf.com).
67. Electrabel is the electricity production leader in the Netherlands. It is part of the French group Suez, an international industrial and service group which carries out its operations in energy and the environment.
68. Enel tried to take control of the French–Belgian company Suez, the third European electricity company. The French government did not want to see one of its industrial jewels change hands to a foreign owner and forced Suez to merge with Gaz de France. The French Constitutional Court, though, put a brake on the operation until the gas market is liberalized, which could happen at the end of 2007.

References

Beder, S. (2005) 'Energía y poder. La lucha por el control de la electricidad en el mundo', Fondo de Cultura Económica, FCE, México. (Original title: Power Play. The Fight for Control of the World's Electricity, 2003, Scribe Publications Pty Ltd.)

Brunekreeft, G. and Twelemann, S. (2004) 'Regulation, Competition and Investment in the German Electricity Market: RegTP and REGTP', Cambridge Working Paper in Economics.

EIA (1996) *Privatization and Globalization of Energy Markets*. Energy Information Administration, Office of Energy Markets and the End Use of the Energy Information Administration, US Department of Energy, Washington. (http://www.eia.doe.gov/emeu/pgem).

European Commission (2001) *Electricity Liberalisation. Indicators in Europe*. DG TREN.

European Commission (2003) *Communication from the Commission on the Energy Infrastructure and Security of Supply*, COM (2003) 743.

European Commission (2004) *Fourth Benchmarking. Annual Report on the Implementation of the Gas and Electricity Internal Market*. COM (2004) 863.

European Commission (2005) *Report on Progress in Creating the Internal Gas and Electricity Market*, COM (2005) 568.

European Commission (2005) *Report on Progress in Creating the Internal Gas and Electricity Market*. Technical Annex to the Report from the Commission to the Council and the European Parliament. SEC (2005).
European Commission (2006) *Prospects for the Internal Gas and Electricity Market*, COM (2006) 841.
European Commission (2006a) *DG Competition Report on Energy Sector Inquiry*, SEC (2006) 1724.
Fabra, J. (2004) 'Liberalización o Regulación? Un mercado para la electricidad', Marcial Pons, Ediciones jurídicas y Sociales, Madrid.
Glachant, J.M. (2003) 'The making of competitive electricity markets in Europe: no single way and no "single market"', in Glachant, J.M. and Finon, D., *Competition in European Electricity Market: A Cross-country Comparison*, Edward Elgar Publishing: Cheltenham, UK.
Green, R., Lorenzoni, A., Perez, Y. and Pollitt, M. (2006) 'Benchmarking electricity liberalisation in Europe', Cambridge Working Paper in Economics, CWPE 0629, March.
Jamasb, T. and Pollitt, M. (2005) *Electricity Market Reform in the European Union: Review of Progress Towards Liberalisation and Integration*, Cambridge Working Paper in Economics, Cambridge MIT Institute.
Joskow, P. (2006) 'Markets for Power in the United States: An Interim Assessment', *Energy Journal* 27(1), 1–36.
Newbery, D.M. (2002) 'Problems of liberalising the electricity industry', *European Economic Review*, 46: 919–27.
Newbery, D.M. (2005) 'Electricity Liberalisation in Britain: the quest for a satisfactory wholesale market design', *Energy Journal*, Special Issue on European Electricity Liberalisation, May: 43–70.
OECD (2004) 'Regulatory reform in Germany. Electricity, gas and pharmacies', in *OECD Review of Regulatory Reform*. Organization for Economic Co-operation and Development.
Perez Arriaga, J.I., Battle, C. and Vazquez, C. (2006) 'Los mercados eléctricos en Europa', in García Delgado, J.L. and Jiménez, J.C. (eds) *Energía: del monopolio al mercado. CNE, diez años en perspectiva*. Thomson Civitas.
REE (2006) *El libro de los 20 años, 1985–2005*, Red Eléctrica Española, Madrid.
Rivero, P. (2006) *La nueva regulación y los retos del futuro en el contexto del mercado único*, Conferencia Jornadas UNESA, 31 May 2006.
Ruiz Molina, M.E. (2003) *Liberalización del Mercado Eléctrico y Elegibilidad: Consecuencias para el consumidor*, Universitat Jaume I.
Thomas, S. (2002) *Why Retail Electricity Competition Is Bad for Small Consumers: British Experience*. Presentation to the International Conference: Restoring Just and Reasonable Electricity Rates, Washington D.C., 28–29 September.
Thomas, S. (2005) *The European Union Gas and Electricity Directives*. Public Services International Research Institute. University of Greenwich.
UNESA (2006) *El sector eléctrico en España, 2006*. Asociación Española de la Industria Eléctrica.

6
Privatizations in Latin America

Gregorio Vidal

Metropolitan University, Mexico

Abstract

Latin America is the region of the developing world where the largest and most important privatizations took place in the 1990s. Argentina, Brazil and Mexico accounted for almost 50 per cent of privatization revenue in developing countries between 1990 and 1999. Nonetheless, this revenue is not significant as a proportion of the region's GDP and public spending. Privatizations are a substantive part of the Washington Consensus proposals. This continues to be the case today: multilateral financial agencies, the United States Treasury Department and Wall Street financial circles insist that privatizations must continue and create new conditions for private investment in infrastructure. In the past, an important part of the resources invested in these activities have been used for acquisitions. Some firms have abandoned the region as a result of other mergers and acquisitions. Still others, however, have been strengthened through purchases in several countries, creating an important degree of economic concentration. The strengthening of several companies and ongoing offers of preferential conditions for investment in infrastructure by some countries have not stimulated significant increases in the capital formation to GDP ratio. Deficiencies in basic public services remain and significant increases in levels of investment in infrastructure are necessary. Given this situation, and along with the growing social inequality and the sheer dimension of poverty in the region, public investment is absolutely necessary.

Keywords: Privatization, Multinational firms, Restructuring, Mergers & Acquisitions, Welfare, State-owned enterprises

Journal of Economics Literature classification: F23, G32, G34, H10, L22 and O1.

Introduction

During the 1990s, of all the developing countries, including the economies in transition, the most numerous and important privatizations took place in Latin America. From the end of the 1980s the sale of public companies and the granting and sale of concessions and licences to private firms to operate different public services became a firmly entrenched part of the region's economic policy. These measures have become a noteworthy, consistent component of the package of economic reforms known as the Washington Agenda or Consensus (Williamson, 1990). In addition to creating a balanced government budget, the Washington consensus considers it absolutely necessary that private companies provide financial services, create and administer infrastructure and other public services to ensure effectiveness and correct resource management.

The premise of private enterprise's intrinsic, superior effectiveness in all manner of economic activities is used to propose and carry out privatization programmes in all regions and countries, whether it is in European Union member countries or sub-Saharan Africa. Today, multilateral financial agencies continue to insist upon it. Other institutions and social actors that are also part of the Washington Consensus are of the same opinion. The World Bank (WB) emphasizes that the private sector must be attracted again, since it has not invested enough to date. The first part of this text reviews the ideas of some multilateral financial agencies, Wall Street institutions and financial circles and other social actors who promote the Washington Consensus agenda.

In some Latin American countries, foreign companies have dominated privatizations; but in others, national capital has made the purchases. Certain privatizations gave rise to private consortia, which held dominant positions in their area of operation, and which quickly became part of the largest companies of their respective countries. Several of them changed hands in subsequent years, becoming part of corporations with home offices outside the region that are now managing public services or infrastructure in several Latin American countries. This is analysed in the second and third parts of this chapter. In addition, as shown by Mexico's government intervention to keep banks operating and repair their financial status, and the support given to some companies in Brazil to deal with their debt in the midst of the crisis of the Real, the state continues to be a key in the process of restructuring large companies important to the national economy. Therefore, privatizations do not imply a simple reduction in the role of the state. They are part of a broader process, the centre of which is increased economic concentration and a distribution of markets among a small group of large corporations, particularly those with headquarters in the United States and some European countries.

Firms define their strategies taking into consideration a world economy that includes blocs and regions, both existing and those in formation. The last part of this chapter argues that this has implied that most new investment is being made in telecommunications. Other kinds of infrastructure currently receive less investment and, as a result, there is a serious lag in this area, even in the region's larger economies. Thus, privatizations have not resolved the problem of infrastructure in Latin American countries and conditions do not seem to exist to ensure that with the participation of a small group of multinational firms important advances can be achieved. Moreover, again as is shown in the last section, privatizations and incentives offered to promote foreign investment in the region and private investment in general have not permitted the gross capital formation to grow as a proportion to GDP. On the contrary, a systemic decrease in this area has been seen in some of the largest economies of the region. In summary, the activity of various foreign companies in the region has not meshed with internal markets, at least not in a way that generates productive chains, deep technological development, and a utilization of the vast and growing productive capabilities, including those of the labour market itself.

Changes in economic management of the state, structural reforms and privatizations

In the mid-1980s International Monetary Fund (IMF)-recommended structural adjustment programmes were being widely applied in Latin America. However, at that time, the region's governments were not contemplating privatizations. A few years before, Mexico had nationalized its commercial banks and the administration had signed agreements with the governments and private corporations of several European countries to create joint investment projects in capital goods industries.

In Western Europe, only the United Kingdom's government had begun to privatize. In 1981 Margaret Thatcher's government sold 50 per cent of the shares of Cable & Wireless, a telecommunications services and installations provider, on the stock market. At that time state-owned companies dominated energy, communications, steel production, naval construction and transportation in the UK. In 1981 the British government privatized its first important company, British Aerospace, the UK's main producer of military and civilian aeroplanes and parts, which had been nationalized in the previous decade. In subsequent years privatizations continued. However, it should be emphasized that in 1979, when the Conservatives won the UK parliamentary elections, their electoral platform included reducing public spending, but they did not propose privatizing public companies (Vidal, 2001). In 1984 the denationalization of the telecommunications network was approved and British Telecommunications PLC was registered as a non-state company allowed trading on the stock market. During the same years, privatizations did not take place in Latin American countries and in other European economies. In Latin America, attempts were made to control inflation through cuts in public spending.

Many companies experienced financial difficulties, were unable to pay their debts and operated with increasingly fragile finances (Minsky, 1986). To understand the diversity of the proposals implemented to deal with low growth and inflation, we should note that in the UK, the state reduced its role as the administrator of public companies and services, while in France and in Greece, actions were taken in the opposite direction. The Mitterrand government carried out a large number of nationalizations as part of a broader proposal that sought to bolster the economy from the demand side with an active role for the state. In Greece, under the PASOK government, which came into power in 1981, the state's role in the economy was also widened. There was no single viewpoint, and governments took diverse initiatives to deal with the economic crisis (Gámir, 1999; De Selys, 1995; Vidal, 2001). However, things changed rapidly and soon a single point of view would come to dominate.

In the United States the Reagan administration took measures to encourage the participation of new capital in public service provision and in the development of infrastructure. American Telephone and Telegraph (AT&T) was dismantled on 1 January 1984 and divided into seven regional companies that maintained the monopoly over local telephone services and an eighth that operated inter-regional and international long-distance service. The long-distance operator kept the name AT&T and made investments abroad. Soon, MCI, GTE-Sprint and US Telephone became its competitors in the US long-distance market. Elsewhere, France Cable et Radio (France Telecom) grew and made investments abroad. AT&T's reorganization, increased investment by several companies that went international and different technological changes modified conditions for accumulation in certain branches of the economy. State action in the economy met with new obstacles, and some business groups found different activities to boost their growth and strength. Since then, a space for significant competition has been created among capital linked to technical development, but also to the trends in foreign investment, privatization of companies providing basic public services (BPS) and the transfer of ownership of assets to foreign firms in several of the most important developed economies and in Latin America, plus more recently in Central Europe and East Asian and Asian Pacific countries.

In June 1992 the World Bank (WB) organized a conference to evaluate the effects of privatizations. Participants included academics and officials in charge of economic policy. A two-year research programme carried out by the WB and Boston University preceded the meeting. Privatizations were considered highly positive and participants insisted on the need to speed up the process world-wide. Lawrence Summers, later US Secretary of the Treasury during the Clinton administration, recognized the difficulties involved in privatizing. He insisted on the need to appropriately combine its execution with other reforms that would reinforce government credibility. He urged initiatives to support companies once privatized including the granting of foreign aid (Galal and Shirley, 1994). In addition to this discourse, the WB, the IMF and the Inter-American Development Bank (IADB) have developed economic policy proposals that indicate a need to advance in privatizations throughout the developing countries and economies in transition. The WB developed its support of infrastructure projects with the participation of private investment, including different forms of technical support. This continues to be one of its most important areas of policy even today.

In Latin American academic circles, privatizations are generally considered to be based on the 'failure of state interventionist policies and import-substitution-based industrialization, like in the stagnation of most of the Latin American countries and the grave budget deficits of their governments' (Cardoso, 1992, p. 8). Those holding these positions maintain that there are studies (Krueger, 1990a and 1990b) that examine the failures of government intervention.

In 1997 the WB wrote its report about the state, maintaining the need to reformulate its role, emphasizing that it cannot be a direct agent of growth. The WB maintains that the reduction of an excessive state presence brings with it two fundamental lessons:

> the liberalization of markets allows new participants to create jobs and wealth. It also lessens the difficulties of privatization.... The second is that, even though an overblown state, should own fewer goods, and while there is no convincing economic reason for industries producing commercial goods to continue to be in state hands, there

is no exact moment in the reform program that can be considered *the most appropriate* to start privatization. (World Bank, 1997, p. 73)

The WB emphasizes the urgency of moving ahead with second generation reforms and changing electoral and party systems, political regimes, the structure of different branches of government and the system of the administration of justice to achieve greater state effectiveness and to ensure that markets act with their natural effectiveness (World Bank, 1997: Chapters 9 and 10). About privatization, it states:

While their importance in the strategy to promote the market may vary from one case to another, many developing countries that want to reduce their oversized state sector must give maximum priority to privatization. A well-managed privatization process produces great economic and fiscal benefits. (World Bank, 1997, p. 7)

As the 1990s advanced the WB maintained its opinion after both the crisis in Mexico, called the Tequila Effect, and the Asian crisis. In a joint study by the investment bank Flemings and the WB on privatizations, the vice-president of finance and development of the WB private sector and Flemings' executive director said that the study was published in the period immediately after the Asian crisis, and for that reason the lessons of the privatization process must be valued even more, particularly in relation to macroeconomic policies and other structural reforms like those of the financial sector so that they can all be successful and permit the development of stock markets in countries that are undergoing privatizations (Rischard and Garrett, 1998, pp. vi–vii). The study goes more deeply into the same tenets and emphasizes the need to keep in mind the lessons of privatizations including the need for the privatization programmes to be transparent, to develop an appropriate legal framework, to establish political support and the measures and institutions that guarantee future competition, and to consider privatizations as an integral part of a whole set of economic reforms (Lieberman and Fergusson, 1998: 6–7). Privatizing, liberalizing financial markets and making reforms to attract foreign investment are a substantive part of the Washington Consensus's structural reform.

After the Tequila Effect, multilateral financial and cooperation agencies recommended that the Mexican government move ahead with privatization. An agenda was established including petrochemicals, airports, railroads, the satellite system and electricity. Something similar happened in Brazil, which accelerated its privatization programme after its economy went through a monetary and exchange rate crisis. Telebrás, several banks and electricity producing and commercialization companies were privatized in 1998 and 1999. Similar recommendations were made for Southeast Asian countries after the crisis in South Korea and the region. An OECD document evaluating privatizations in member countries stated that privatization is a central component in the tasks of government reform, emphasizing the 2001 sale of part of Korea Telecom's capital (OECD, 2002). Even during Argentina's deepening economic recession at the beginning of the current decade, international financial organizations maintained the position that privatizations must continue.

More recent documents published by multilateral financial agencies like the WB and the IADB continues to maintain the need to deepen the privatization process. The increase in private companies' operations of the public services and the region's relative importance among developing countries in the sale of public companies until the end of the 1990s are held up as advances. In 2001 the IADB published a study on structural reforms in Latin America. The study's author, Eduardo Lora, suggests measuring their advance using the *index of structural policies* that includes trade, financial, tax, privatization and labour policies. With regard to privatizations, he maintains that countries in the region made noteworthy, if irregular, progress. 'The 396 sales and transfers to the private sector effected in Latin America between 1986 and 1999 represent more than half the value of the privatizations in all developing countries, with the exception of those done by massively distributing coupons in the Eastern European countries' (Lora, 2001, p. 16). In the 1990s the index created based on the accumulated value of sales and transfers of companies as a proportion of the GDP during the corresponding year showed a notable increase, particularly in Bolivia, Peru, Brazil and Argentina (Lora, 2001, p. 26). According to this study, completing the privatizations is a necessary accompaniment to the entire programme of structural reforms.

The World Bank has published several studies emphasizing the advances in private participation in infrastructure. However, the WB maintains that in some activities, investments by private companies are still insufficient. As such, there are some public services that still should be privatized among a large group of companies. In the early 1990s private companies provided only 3 per cent of telephone connections and electricity in the region and almost no water services. By 2003 private companies were providing 86 per cent of telephone service and 60 and 11 per cent of electricity and water hook-ups, respectively (Andrés *et al.*, 2005). The private sector's advance in telephone services is particularly noteworthy. But a broad field remains open to privatizations in the generation, distribution and commercialization of electricity. From the WB study's perspective, even more remains to be privatized in water, sanitation and transportation.

When evaluating the results of privatizations, analysts reiterate the reasons they were necessary. An IADB study published in 2003 based on a vast bibliography on the issue insists that state-owned enterprises (SOEs) operate with flawed monitoring and poor incentives for their boards of directors, resulting in performance inferior to that of private companies. Among the reasons for this behaviour are: the average state company is not registered on the stock market and is not subject to buy-outs by other firms. They are also not subject to the discipline that a creditor establishes because they are financed with public debt and their losses can be covered with public funds (Chong and López de Silanes, 2003, p. 7). These authors and others add that SOE boards of directors rarely exercise good corporate governing practices and their leadership is subject to more political reasons than to the forces of the market (Vickers and Yarrow, 1988: Chong and López de Silanes, 2003).

As pointed out earlier, for the forces of the Washington Consensus the market is effective in enforcing efficiency production. But, in addition, those who think that boards of directors act in their own interests only in the case of SOEs are wrong. This is a generalized problem in large firms. This is why new, diverse regulations are drawn up, measures for incorporating independent council members are established, watchdog bodies impose sanctions and board members in large corporations are removed from their posts for irregularities or even brought up on criminal charges. Despite all of this, companies continue to go bankrupt and suffer from bad management both in the United States and in Europe and Japan. Bankruptcies are carried out according to laws that permit different kinds of government support, including tax subsidies. Or, as has happened in several countries in Latin America, some of the privatized companies that went bankrupt or suffered severe financial problems were bailed out with public funds.

The defenders of the Washington Consensus agenda argue that in SOEs political issues dictate company policy, resulting in an excessively large work force, a poor selection of products and location and inefficient investment (Schleifer and Vishny, 1996; La Porta and López-de-Silanes, 1999; Chong and López de Silanes, 2003). However, this does not add a new argument. In private firms, boards of directors also defend their own interests, resulting in many lay-offs, plant closures and firms relocating from one country to another. And these decisions do not necessarily take into account stockholders' interests. This can be read as the result of bad prior decisions or as part of a profound restructuring. The phenomenon is not limited to a particular kind of company. Conflict of interests between managers and stockholders is widely recognized as a general issue; private enterprise does not eliminate it, but rather functions under this constant tension between owners and managers.

Thus, regardless of repeated crises in different developing countries, government intervention to support or to take over the management of certain recently privatized companies and the continual changes in ownership of many companies, the institutions that support the Washington Consensus conclude that it is necessary to continue privatizing. This is the firm recommendation of international financial agencies, Wall Street financial circles, and the boards of directors of multinational corporations. Their postulate is simple: private enterprise is simply more economically effective than state companies. However, the facts are more complicated.

Regional and sectoral concentration in privatizations

In Latin America privatization processes were preceded in many countries by increased state involvement in the economy and nationalizations. However, since the end of the 1980s, steel, telecommunications, banking, electricity and airlines all began to be privatized. Later, many public services were licensed out to private firms, including a growing number of foreign companies that only a few years before had begun internationalizing. In some countries, changes in the structure of production and ownership were made in different branches of the economy. However, this process was not continuous. In some countries it slowed down for several years. There was also considerable difference in timing between one country and another, even though they all said they were applying economic policies based on similar principles.

From 1990 to 2005 US$203 billion were generated in Latin American countries from privatizations. In 1988 and 1989 the Salinas de Gortari administration launched its broad privatization programme. Meanwhile, in Argentina, Menem's government presented its privatization programme in late 1989, advancing rapidly in 1990 and 1991. In Brazil, Collor de Melo announced his national privatization programme in the early 1990s. Most privatizations, both in terms of number and worth, were concentrated in the decade from 1990 to 2000, making up 96.4 per cent of these operations in the period ending in 2005.

As can be observed in Table 6.1, in terms of their worth, privatizations intensified in the last third of the last decade, from 1997 to 1999

Table 6.1 Latin America privatization revenues, 1990–2005 (billions of dollars)

	Mexico	Argentina	Brazil	Peru	Venezuela	Chile	Total
1990	3.11	7.53	0.04		0.01	0.10	10.92
1991	11.29	2.84	1.63	0.003	2.28	0.36	18.72
1992	6.92	5.74	2.40	0.21	0.14	0.01	15.56
1993	2.12	4.67	2.62	0.13	0.04	0.11	10.49
1994	0.77	0.89	2.10	2.84	0.01	0.13	8.20
1995	0.17	1.21	0.99	1.28	0.04	0.01	4.62
1996	1.53	0.64	5.77	1.77	2.02	0.19	14.14
1997	4.50	4.37	18.74	1.27	1.39		33.90
1998	1.00	0.51	32.43	0.48	0.11	0.18	37.69
1999	0.29	16.16	4.40	0.29	0.05	1.05	23.61
2000	0.41	0.03	10.70	0.38	0.02	0.43	18.46
2001		0.01	2.57			0.24	3.14
2002			0.004	0.28			0.30
2003						0.31	0.41
2004			0.65	0.40		0.78	2.19
2005	0.35		0.07				0.92

Sources: World Bank, Global Development Finance, 2000, p. 138, 2001, p. 186; World Bank, Privatization Database http://rru.worldbank.org/Privatization/Region.aspx?regionid=435, May, 2007; Governo de Brasil, Ministério do Desenvolvimiento, Indústria e Comércio Exterior, Privatização no Privatização no Brasil, 2002; ECLAC, Informe sobre la Inversión Extranjera en América Latina y el Caribe, 2001; SELA, El financiamiento externo y la deuda externa de América Latina y el Caribe, December 2002. OECD, Recent Privatization Trends in OECD Countries, No. 82, June 2002.

representing 46.8 per cent of the income generated for the entire period of 1990 to 2005. Infrastructure tops the list, followed by the privatization of pension funds, the sale or licensing to the private sector of certain health services and different experiments with managing education with private funds or according to the exclusive needs of certain groups of companies. The difficult economic situation of 1998 did not slow down structural reform (ECLAC, 1998). In 1998 Latin America's governments received US$37 billion from privatizations (see Table 6.1), including transactions in the smaller economies like those of Central America and involving activities such as mail delivery, though most (86 per cent) were concentrated in Brazil (ECLAC, 1998, p. 6).

Brazil, Argentina and Mexico accounted for the largest amounts of revenue from the sale of public companies: together, they come to 79.8 per cent of all the cumulative revenue (see Table 6.1), the result mainly of the sale of a small number of banks and companies providing basic public services. The largest revenues came in Mexico in the early 1990s and in Brazil in the latter half of the decade. In Brazil privatizations were suspended with the fall of the Collor de Melo administration, while in Argentina, they continued apace. According to annual figures, privatizations were only important in Venezuela in 1991 and 1996 (see Table 6.1). However, privatizations did not continue there in the following years, at least in the petroleum sector. The Chilean economy's case is different given that important privatizations had taken place there since the 1970s, although the figures rose again in 1999 and 2000, when public services began to be privatized again.

In 1996, in addition to Venezuela, Colombia registered significant revenues from privatizations. Jointly, the two countries' revenue from this source represented, 28.5 per cent Latin America's total privatization revenue for the year. In 1996 Colombia's income from this source totalled US$2 billion; this, plus that garnered in 1997, amounts to 68 per cent of income from privatizations from 1990 to 2005. As in the case of other medium-sized and small economies in the region, it was the sale of electricity generation and distribution companies or phone companies that accounted for most of the revenue received. In 1998 Brazil's large income came primarily from the privatization of its phone system and the licensing of mobile phone services. In 1999 Argentina concluded the sale of Yacimientos Petrolíferos Fiscales (YPF) stock, and in 2000 and 2001 Brazil began privatizing some of its state banks. The companies privatized and the acquiring firms are listed in Table A.6 of the Appendix. Mexico proceeded to privatize its airports, and a first discussion began about the sale of electricity companies while more private producers of electricity were allowed to invest.

Argentina, Brazil and Mexico account for the majority of assets privatized in the region. For the acquiring companies, reaching a dominant position in these three economies was key to their regional strategies. Several companies sought a regional foothold in the region. This objective required an important position in these three countries. Brazil, Mexico and Argentina account for almost 50 per cent of the income generated from privatizations from 1990 to 1999 in developing countries, including Eastern Europe, Central Asia and Eastern Asia and the Asian Pacific (see Figure 6.1). In that same period, no other economy in the developing world has the same weight. China contributes barely 7 per cent, and India, which has a large number of public companies, 3 per cent. Between 1990 and 2001, South Korea, which the WB and other multilateral body reports consider a high-income country, received a greatly inferior amount from privatizations than any of the three largest economies of Latin America: US$14.891 billion, 34 per cent of which corresponds to 1999 to 2001 (OECD, 2002, p. 46).

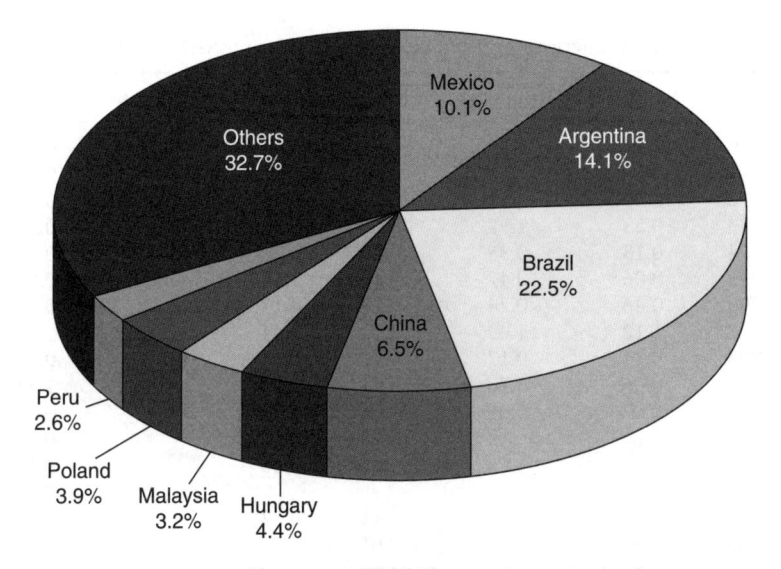

Figure 6.1 Privatizations in developing countries, 1990–99
Source: World Bank, Global Development Finance, 2000 and 2001.

By sector, sales of companies have been concentrated in telecommunications and electricity. As mentioned above, in almost all cases, the largest sums accrued from privatizations have come from the sale of telephone companies and, in some countries, companies in the electricity or petroleum sectors. In Colombia and Venezuela the electric companies particularly stand out, while in Brazil and Argentina petroleum predominates. Privatizations of telephone companies and licensing to develop mobile phone services were all undertaken during the 1990s. Electricity companies were also sold off. Given that no other activities require large amounts of resources for purchasing state assets or that it has not been possible to privatize some electricity companies, the region has lost importance among developing countries.

During the 1990s privatizations in Latin America were important because they allowed some companies to further their internationalization. As the following sections analyse, the result has not been the strengthening of companies with headquarters in the region, with the exception of Telmex-América Móvil-Carso, headquartered in Mexico, and more recently Petrobrás, with its main offices in Rio de Janeiro (still partly state-owned). However, at the economy-wide level, in terms of the size of GDP and public budgets, income from privatizations has not been significant. Among the group of larger regional economies, revenue

Table 6.2 Latin America countries: privatization revenues/GDP, 1990–2005 (%)

	Mexico	Argentina	Brazil	Peru	Venezuela	Chile
1990	1.20	5.80	0.01		0.02	0.29
1991	3.59	1.63	0.40	0.01	4.40	0.95
1992	1.90	2.73	0.61	0.59	0.24	0.02
1993	0.53	1.97	0.60	0.36	0.06	0.22
1994	0.18	0.35	0.39	6.32	0.01	0.23
1995	0.06	0.47	0.14	2.38	0.05	0.02
1996	0.46	0.24	0.74	3.13	2.95	0.25
1997	1.12	1.49	2.32	2.14	1.62	
1998	0.24	0.17	4.12	0.85	0.12	0.23
1999	0.06	5.70	0.82	0.55	0.05	1.44
2000	0.07	0.01	1.78	0.71	0.02	0.57
2001			0.50			0.34
2002				0.46		
2003						0.42
2004			0.11	0.57		0.82
2005	0.05		0.01			

Source: Author's elaboration based on Table 6.1 and ECLAC, *Anuario Estadístico de América Latina y el Caribe 2006* in www.cepal.org. October, 2007.

obtained from privatizations has tended to be less than one per cent of GDP per year. As can be seen in Table 6.2, this is the case with Mexico, with the exception of 1990 to 1992 and 1997. The year 1991 stands out as the year in which the telephone company was privatized along with the majority of the 18 commercial banks then in government hands. These exceptional revenues did not modify the structure of public spending. Brazil presents a similar case. In 1998, when privatization revenues equalled 4.12 per cent of GDP (see Table 6.2) a large part corresponded to the sale of the telephone system and to the concessions sold for the provision of cellular telephony.

In Argentina, the high levels of revenue as a proportion of GDP in 1999 were the result of the conclusion of YPF's privatization. In Peru the year 1996 can be explained primarily by the sale of the state telephone company. Discounting the years in which the greatest amounts of revenue were recorded, the remaining privatization revenue is irrelevant in relation to public expenditures. As public spending is on average equivalent to 26 per cent of GDP per year, revenues equal to or less than one per cent of GDP do not imply modifications for public finances. In addition, as can be seen in Table 6.2, since 2000 privatization income has been greatly reduced, in cases where it is still registered.

Confirmed through the comparison with other macroeconomic figures is the fact that although states of the region sold assets, of which some produced consistent income, they did not gain improved conditions for public expenditure in return. Moreover, if one considers that foreign capital entered the region by way of privatizations, it is also true that standard relations between countries of the region and foreign companies and transnational financial institutions imply the outflow of much greater amounts of resources in just a few years. The total net payment of interest and earnings that countries of the region sent abroad is a larger figure than the amount of income obtained through the sale of state assets. In the period of 2000 to 2005 the balance of income registered a deficit of 361 billion dollars, equivalent to 177 per cent of privatization revenues from 1990 to 2005. The funds obtained through privatizations from 1990 to 2005 were equivalent to 8 per cent of the region's GDP in dollars in 2005. Although the distribution is uneven among countries, the greatest earner did not exceed 16 per cent. Revenues earned between 1990 and 2000 as a proportion of 1999's GDP were 11.5 per cent for Argentina, 9.9 per cent for Brazil, and 7.5 per cent for Mexico (Lora, 2001). For the entire period of 1990 to 2005, privatization revenue represented only 0.6 of Latin America's GDP.

Beginning in 2001, the highest amounts accrued from privatizations were to be found in Eastern Europe and China: China accounted for 20 per cent of the total and Poland 8 per cent. The Czech Republic, Turkey and Slovakia also reported significant amounts. In the entire period the three largest Latin American economies received 35.9 per cent of income from privatizations, putting them among the five economies that have received the most between all the developing and emerging countries. From 2000 on, Latin America saw important changes in ownership of privatized assets, and several multinational corporations that had purchased state companies sold their interests, as did some consortia in the region. Property ownership transfers were very intense in the telephone sector. But these were also years in which the problems of growing inequality and high poverty levels highlighted the need for major investments in some public services.

Foreign capital and privatizations

During the 1990s and until 2001 the Latin American governments that received the most income from privatizations were those same countries that were foremost in foreign direct investment (FDI). Mexico (from 1991 to 1994) and Brazil (in most years after 1995) have been the two

Table 6.3 Brazil's ten largest privatizations with foreign investors, 1987–2005

Privatized firm	Sector	Year	Acquiring firm	Acquisition (billions of dollars)
Telesp Participacoes S.A.	Telecommunication	1998	Telebrasil Sul Participacoes S.A. Consortium: Telefonica S.A. and Subsidiaries (56.6%), Telecom de Portugal (23%), Iberdrola Investimentos (7%), BBV(7%), RBS Participacoes (6.4%); Spain (70.6%); Portugal (23%), Brazil (6.4%)	5
Petrobras	Petroleum	2000	Partial sale. Included ADRs and equities in Sao Paulo and Madrid	4.2
Banco do Estado de São Paulo (BANESPA)	Bank	2000	Banco Santander Central Hispano, Spain	3.55
Telesp Celular	Telecommunication	1998	Telecom Portugal (100%)	3.1
Light Servicos de Electricidade	Power Utility	1996	AES, USA(20.3%); Houston, USA (20.3%), EDF, France (20.3%); BNDES, Brazil (16.4%)	2.3
Embratel	Telecommunication	1998	MCI (100%)	2.3
Telecentro Sul	Telecommunication	1998	Techold Part, Italy (19%)	1.8
COELBA	Power utility	1997	Iberdrola, Spain (39%)	1.8
Elektro	Power utility	1998	Enron, USA (100%)	1.7
Cia de Electricidade do Estado da Bahia	Power utility	1997	Iberdrola (39%)	1.6

Source: Author's elaboration based on UNCTAD, World Investment Report 2000, United Nations, Ginebra, 2001, p. 134; BNDES, Privatizações no Brasil 1990–2004, Rio de Janeiro, February, 2005. ECLAC, *La inversión extranjera en América Latina y el Caribe*, 2002.

Table 6.4 Argentina's largest privatizations with foreign investors, 1987–2005

Privatizated firm	Sector	Year	Acquiring firm	Acquisition (billions of dollars)
YPF S.A.	Petroleum and Gas	1999	Repsol, Spain (84%)	13.2
Aeropuertos Argentinos	Airports	1998	Empresas, USA	5.1
YPF S.A.	Petroleum	1999	Repsol, Spain (14.9%)	2.0
Telefónica de Argentina	Telecommunication	1990	Telefonica, Spain (10%); Citibank, USA (20%) Others foreign banks (37%)	2.0
Telecom Argentina	Telecommunication	1990	Telecom Italy (32.5); France Telecom (32.5%)	1.8
Aerolíneas Argentinas	Airline	1990	Iberia Airlines, Spain (69.4%)	1.3

Source: Own elaboration supported by UNCTAD, World Investment Report 2000, United Nations, Ginebra, 2001, p. 134; BNDES, Privatizações no Brasil 1990–2004, Rio de Janeiro, February, 2005. ECLAC, La inversión extranjera en América Latina y el Caribe, 2002.

economies that received the largest amounts of capital from abroad (Vidal, 2001: Chapter 2). Sales of state assets had an impact on stock market growth, both in the countries of origin and in the US. Some of the most important privatizations involved floating stocks and American Depository Receipts on the New York Stock Exchange. Others took place after changes to legislation to attract capital from abroad. However, in 1989 and the first part of the 1990s, Mexican and Brazilian companies being privatized were purchased by national capital. This was partly because of the conditions of sale. Nothing similar occurred in the other economies, including Argentina.

Other financial corporations or national investors purchased Mexico's banks. In the case of Teléfonos de México, control of the firm ended up in the hands of the Carso group and domestic investors, with participation by Southwestern Bell International Holdings and France Cables et Radio. In addition, Telmex stock was sold on different stock markets throughout the world. In Brazil, the Usiminas and CSN steel works, national investors purchased the two biggest companies privatized from 1990 to 1995. The privatization of the Compañía Vale do Rio Doce (CVRD) mining corporation took place in three stages: in the first, the firm's control package was sold to a group of investors with foreign participation, but organized with Brazilian capital. The Brazilian government privatized other companies selling part of the stocks on international stock markets, and another part on the country's stock exchange. The result, as Table 6.3 shows, is that in these years, foreign firms acquired no important privatized companies. By contrast, in Argentina the controlling interest in the telephone system and Aerolíneas Argentinas was purchased by foreign firms (see Table 6.4). Foreign investors purchased 75 per cent of YPF (IADB, 1996, p. 182). The government later sold its remaining 25 per cent share in the company. In both cases part of the capital was sold on the stock market.

In Peru and Bolivia foreign corporations purchased more than 80 per cent of the value of companies being privatized (IADB, 1998, p. 185). In Venezuela, from 1990 to 1997, foreign investors purchased 80 per cent of all assets privatized. In 2000, when Electricidad de Caracas was privatized, US-based AES Corporation acquired 81.3 per cent of its assets. In other regions' economies, the resources amassed came almost exclusively from foreign direct investment in direct sales (see Appendix, Table A6). The IADB emphasizes that: 'For example, 100 per cent of all income from privatizations in the Caribbean and Central American countries has been FDI' (IADB, 1996, p. 185).

As the 1990s wore on privatizations continued and the participation of US- and European-headquartered companies increased. Brazil sold its electricity companies, its gas distributors, and, notably, Telebrás; in all these cases, foreign firms participated, in some cases, important multinational corporations, as major partners (see Table 6.3).

In short, as Table 6.5 shows, foreign firms headquartered outside of Latin America acquired a series of privatized companies that belong to specific sectors. The most noteworthy cases are telecommunications and electricity. In both these sectors purchasing firms had only been recently established. Several of them began internationalizing in the 1990s or at the earliest during the late 1980s. Most noteworthy is that privatized steel, mining and television companies remained in national hands even though long-established multinational corporations did exist.

Outstanding among the companies that began internationalizing with acquisitions in Latin America are telephone, electricity, gas and petroleum companies, banks and financial services institutions. Noteworthy in this sense are the Spanish companies Endesa, Iberdrola, Unión Fenosa, Repsol, Telefónica, BBVA and BSCH. Significantly, some US-based companies also internationalized in this way: MCI, Enron, AES and BellSouth. It is also worth noting that Japanese companies did not participate at all, and in general Japan has purchased few assets in the region.

In the context of the Argentinean crisis, there were two new developments, one in Brazil and the other in Argentina. In Brazil banks

Table 6.5 Highlights in Latin American privatizations

Sector	Period	Buyers	Main names
Airlines	1988–90 and 2001	Domestic and foreign from Spain	Iberia
Telecommunication	1990 and 1998	Domestic and foreign from Italy, US, France and Spain	Telefonica, MCI, Bell South, Verizon, France Telecom and Telecom Italy
Power Utility	1990–2000	Foreign from Spain, France, US, Canada, Portugal	Repsol, Endesa, Iberdrola, Enron, AES, Electricite de France
Gas	1994–2000	Foreign from Spain, UK	Enron, Iberdrola
Railroad	1996–1997	Domestic and foreign from US	Kansas City Southern Industries and Union Pacific
Airport-Infrastructure	2001 and 2005	Foreign from Spain	
All alone 1990–2000			
Petroleum		Domestic and foreign from UK, US, France, Spain	Repsol, Total, British Petroleum
Financial		Domestic and foreign from Spain, US	Santander y Met Life
Mainly domestic			
Steel	1991–94	Domestic	
Mining	1988–90 and 1997	Domestic	
Broadcasting	1990–93	Domestic	

Source: Author's elaboration based on World Bank, Privatization Database 2000 and Privatization Database 2000–05, http://rru.worldbank.org.

and bank assets continued to be purchased, but the buyers were banks headquartered inside the country, particularly Bradesco and Itaú, the nation's largest private banks. They have even acquired banks that had previously been in foreign hands. This is part of the maintenance and growth of some Brazil-based entrepreneurial groups, which have made acquisitions in diverse sectors and have achieved a certain level of internationalization. In 2002 Brazilian-based corporations made 27 of the largest acquisitions in Latin America, a large number of which were of Brazilian companies. In addition to banks they purchased breweries, steel plants, mines, paper and cellulose plants, petroleum, petrochemicals and other chemical industries. Among these entrepreneurial groups are Gerdau, CVRD, CSN, Votorantim, AmBev and Petrobrás (LatinTrade, 2003, pp. 40–1). These were not acquisitions of state companies but of private firms, some owned by multinationals. The important point is that this showed the relative strengthening of some business groups in Brazil. At that time Brazil's privatization programme was on hold, the first step toward it being eliminated at the beginning of Lula's first term.

In Argentina, a series of foreign companies' subsidiaries closed down, several of which had participated in privatizations in previous years. Canada's Bank of Nova Scotia pulled out and the branches of its affiliate Scotiabank Quilmes were distributed between two nationally owned banks, Bancosud-Macro and Banco Comafi. France's Crédit Agricole also withdrew and Italy's Intesa BCI became the minority partner in the Banco Patagonia (ECLAC, 2003, pp. 35–6). Other foreign banks with subsidiaries in Argentina, such as Spain's BBVA and BSCH, announced they would not put money into their subsidiaries in Argentina. The owners of public utilities announced they were suspending investment and began demanding rate hikes and special tax breaks. Some companies had difficulties in issuing debt abroad and, contrary to what Argentine government authorities claimed when the privatizations were undertaken, the headquarters of these companies did not come to the aid of their subsidiaries. As discussed in the final section, the companies that experienced financial difficulties, declared moratoriums on their debts, or left the country had operated in those sectors that remitted the most profits abroad in the 1990s. This is an important indicator of the economic result of privatizations in Argentina and the role of foreign capital itself.

Some companies also left the region because of changes in ownership of their home offices or in order to concentrate investment in other markets. Some companies are experiencing serious financial difficulties and others, like Enron or MCI, are in bankruptcy or undergoing complete restructuring, leading them to sell their assets in Latin America.

AT&T, BellSouth, MCI/WorldCom and GTE, all headquartered in the United States, made important investments in telephone systems. With the exception of AT&T, these companies are rather new. Recently they have undergone several mergers and have sold their assets in the region. Verizon came into play later as a result of mergers, but Telecom Italia and France Telecom have also made important investments. However, despite initial large investments, as Table 6.6 shows, with the passing of years several of these firms sold their shares, while investments by Telefónica de España and Mexico's Telmex-América Móvil grew.

Telephone services are important for understanding the role of privatization and the extent of private investment in creating infrastructure.

Table 6.6 Latin America: private investment in strategic and basic public services Share of the largest 19 firms in total private investment: 1990–2002

Firm	Country of origin	Sector (1)	Public assets sold	New investments	Total investments	Share of total investment (2) (%)
Telefónica	Spain	Telecomunication	11.29	23.34	34.64	9.81
Carso/Telmex/América Móvil	Mexico	Telecomunication	9.90	20.77	30.68	8.68
Telecom Italia	Italy	Telecomunication	7.88	17.71	25.59	7.24
France Telecom	France	Telecomunication	2.42	9.57	11.98	3.39
Verizon	United States	Telecomunication	3.51	9.16	12.67	3.59
BellSouth	United States	Telecomunication	3.97	5.79	9.76	2.76
MCI/WorldCom	United States	Telecomunication	1.86	5.27	7.13	2.02
Endesa/Enersis	Spain	Electricity	8.32	9.34	17.65	5.00
AES	United States	Electricity	10.33	5.66	15.99	4.53
Enron	United States	Electricity	3.66	8.30	11.86	3.36
Électricité de France	France	Electricity	3.67	3.18	6.85	1.94
Iberdrola	Spain	Electricity	4.85	3.20	8.06	2.28
Aguas de Barcelona	Spain	Water and sewage	1.30	7.01	8.31	2.35
Perez Companc/Petrobras	Brazil	Gas and electricity	3.33	3.64	6.97	1.97
Others (five firms)			19.71	38.82	58.63	16.60
Total (group of 19 largest)			96.00	170.77	266.77	75.52
Total private investment			116.60	236.66	353.26	100.00

Notes: (1) This refers to the activity that constitutes more than 80% of the firm's total investment in basic or strategic public services. (2) This is the company's share of all private sector investment in basic or strategic public services in Latin America
Source: Author's elaboration based on: World Bank, Private participation in infrastructure project database, http://rru.worldbank.org/ppi/reports/

This is a branch of the economy in which investments are linked to new technology (mobile phone service), but are also significantly concentrated. In all of the region's economies except Brazil, one company dominates landline services that handle the communications network and, together with one other firm, dominates mobile service. In some of the economies, the dominant company has a more than a 60- or 70-per cent share of the market, a level of economic concentration that has been created after several sales of assets. The largest US, Spanish, French and Italian companies have carved out places for themselves in the region, as did Britain's largest mobile phone group. However, the countries' communications needs have not been covered. The index of basic telephone lines in the region is just a third of that of higher income OECD countries (UNDP, 2006). As will be shown later, there has been a greater penetration in cellular telephony. However, as evidenced by the existence of regions still lacking basic telephone services, these gains have been superimposed on existing conditions of social inequity. The problem is even more pressing in other areas of infrastructure, such as potable water and drainage services (UNDP, 2006).

Private Capital in Basic Public Services and Multinationals

Between 1990 and 2002, private investment in basic public services (BPS) came to US$346 billion, which went into telecommunications (47.1 per cent), electricity (25.5 per cent) and highways (8.3 per cent). The rest was distributed among the transmission and distribution of natural gas (5.5 per cent), water and sewage (5.1 per cent), railroads (4.7 per cent), airports (2 per cent) and seaports (1.8 per cent). Of all the resources invested, 33.6 per cent correspond to privatizations and the rest are new investments by private capital. From 2003 to 2005, no important changes occurred in the areas targeted for investment, but telecommunications benefited from an increase in resources that came from property changing hands. In that same period, investment in infrastructure including private participation came to US$42 billion, of which 79 per cent went into telecommunications. Much of this investment was made by a small group of companies.

In 1998 the private sector spent US$23.885 billion to purchase government telecommunications assets, 45.4 per cent of all expenditures in purchasing government assets in the sector from 1990 to 2002. In 1997 and 1998 it purchased a full 55 per cent of what it bought overall. In those same years FDI income was much more than the average for the 1990–95 period, with the purchase of public service providers constituting one of the most important items these funds were earmarked for. From 1997 to

1999 Brazil received more than US$71 billion in FDI. But just the privatization of phone service and the bidding for mobile phone service carried out in their immense majority between 1997 and 1999 brought in more than US$25 billion. Brazilian capital did participate in the purchase of the telephone companies, but, as was pointed out earlier, an important part of the funds were FDI (see Table 6.3). In addition, in the following years, the participation of foreign partners increased.

In other years when private participation in the purchase of BPS was high, FDI was also high. However, this was the case in other economies. In 1991 and 1992 the outstanding case is Argentina, which privatized its railroads, gas distribution network, electricity utilities and phone service. Undoubtedly this is the Latin American country that at that time began to implement the most important programme of transferring BPS to private capital. FDI played a part in several companies, particularly from companies headquartered in the European Union with majority state ownership.

In Mexico, the process was different. There were important privatizations from 1990 to 1992 but only the telephone company was a BPS. Electricity, the distribution of natural gas, railroads and ports were not privatized. It was not until 1995 that private capital was able to buy into ports, and until 1996 into railroads. But, in addition, the phone company was privatized without splitting it up, and certain rules were set up to allow most voting capital to remain in the hands of the Mexican partner. This was the only case of all the most important Latin American economies in which this kind of a procedure was established. In Venezuela, the privatization of CANTV gave foreign capital a share in the firm, as was the case in Peru.

Thus, in Mexico, private capital participation in BPS, with the exception of the phone service, did not happen until the second half of the 1990s, when all restrictions on FDI were lifted. In Venezuela, after the privatization of the telephone service, which accounted for more than 90 per cent of the income derived from the sale of BPS, the electric company, Electricidad de Caracas, was sold to AES, as pointed out earlier. The sale of Peru's two telephone companies, CPT and ENTEL, in several stages was also very important. Initially, Telefónica de España put up 35 per cent of the capital for the sale, later increasing its share to 97 per cent. Electricity generating and distribution companies were also sold with no restriction whatsoever of foreign capital participation.

Thus, in the second half of the 1990s, when FDI in the region was growing and there was a large number of cross-border mergers and acquisitions, privatizations increased and companies that had been operating for only a short time with some private capital ownership were sold off. Several firms made purchases in different Latin American countries. As mentioned in the preceding section, outstanding purchasers in telecommunications included Telefónica de España, Telecom Italia, France Telecom, the US companies Verizon, BellSouth and WorldCom-MCI and Mexico's Grupo Carso, which included Telmex, which is separate from América Móvil. The United States' AES and Enron, Spain's Endesa-Enersis and Iberdrola and Électricité de France made transactions in electricity. Enron and Iberdrola also own important natural gas distributing companies. Finally, the largest consortium in Latin America in potable water and drainage is Aguas de Barcelona. Table 6.6 shows these companies' investments in basic public services in the years 1990 to 2002. The last column shows each company's participation as a percentage of all private capital investment in BPS, including purchases of government assets. Of the 14 companies listed, 12 are headquartered outside Latin America, and by 2002 accounted for 48.3 per cent of private investment in BPSs. The five largest, all of them in telephone services, accounted for 32.7 per cent of these investments. They also made 34 per cent of new investments, but this includes different asset purchases after privatization. This phenomenon is particularly important in telephone services and other BPSs.

Iberdrola withdrew from Brazilian telecommunications when Telefónica launched a bid for 100 per cent of Telesp, Tele Sudeste and Tele Leste in 2000. Telefónica's participation in Venezuela's Verizon-controlled CANTV also changed. And more recently it changed again with the withdrawal of foreign firms when the Venezuelan government acquired it. France Telecom unloaded its shares in Mexico's Telmex, and, in October 2003, sold its shares in El Salvador's CTE to América Móvil. Verizon left the Argentinean market, and MCI sold its part of Brazil's Embratel to América Móvil. AT&T sold its Latin American holdings to América Móvil, and BellSouth to Telefónica.

Outstanding among recent acquisitions of electricity and natural gas companies are Petrobras's purchase of a majority share in Perez Companc, headquartered in Argentina, which also deals in petroleum. Perez Companc was among the 19 largest companies with investments in BPSs, and this acquisition makes it part of the Petrobras group. The growth of América Móvil, a split-off from Carso Global Telecomunicaciones, also involves the purchase of several mobile phone firms in Latin America and the Caribbean.

Largely as a result of acquisitions, and not necessarily improving the region's infrastructure, by 2005 the list of companies that had made the

Table 6.7 Latin America: private investment in public services
Largest 12 firms in total private investment: 1990–2005 (billion dollars)

Firm	Origin Country	Sector	Public assets sold	New investment	Total investment
Carso/Telmex/ America Movil (1)	Mexico	Telecomunication	16.81	40.64	57.45
Telefónica	Spain	Telecomunication	16.90	37.29	54.19
Telecom Italia (2)	Italy	Telecomunication	8.63	25.12	33.75
Suez (3)	France	Energy, water and sewage	2.15	11.77	13.92
AES Corporation	United States	Electricity	8.52	4.75	13.27
Endesa/Enersis	Spain	Electricity	5.28	5.12	10.40
Verizon	United States	Telecommunication	3.18	6.32	9.50
Enron	United States	Electricity	1.91	7.31	9.22
Aguas de Barcelona	Spain	Water and sewage	1.71	7.45	9.16
Iberdrola	Spain	Electricity	4.06	4.77	8.83
Electricite de France	France	Electricity	3.99	3.17	7.16
Petrobas	Brazil	Energy	2.51	3.53	6.04

Notes: (1) Carso invierte en construcción de carreteras: 730 millones de dólares (322 million dollars in public assets sold and 408 million dollars in new investments). (2) Telecom Italia: investments with Grupo Werthein, Argentine (Telecom Argentina); con Techold Participacoes, Brazil (Brasil Telecom); con ABC Telecomunicaciones, Paraguay (Telecom Personal). (3) 6.13 billion dollars in energy, 7.52 billion dollars in water and sewage and 0.28 billion dollars in transport.
Source: Author's elaboration based on: World Bank, Private participation in infrastructure project database, http://rru.worldbank.org/ppi/reports/

largest investments in BPS changed considerably. As Table 6.7 shows, the five largest include one energy firm and another focused mainly on water and sanitation. All together they have made 42.5 per cent of the investments between 1990 and 2005. It should be noted that France Telecom is no longer one of the largest investors since it unloaded its holdings. BellSouth and MCI/WorldCom are in the same situation. Looking at sales in 2006 and 2007, Verizon and probably Enron are also no longer among the largest investors. This is all the result of mergers and acquisitions, but explains a large part of the new investments.

The changes in stock holdings, some determined by acquisitions or mergers in the corporations' home countries and others by acquisitions in markets outside Latin America and the Caribbean have drastically altered business structures and their growth strategies. By 1998 the region's telecom sector hosted four of the largest US headquartered companies: AT&T, BellSouth, MCI and Verizon; the largest cellular phone company in Europe: Verizon; the dominant company in France, also with investments in other European countries: France Telecom; the largest Italian Company: Telecom Italia; and the most important Spanish company: Telefónica. All of these had purchased privatized companies in a large number of Latin American countries. Telefónica remains in the region, competing with Carso/Telmex/América Móvil in various countries. The third operator with important investments in Brazil is Telecom Italia, although it was the object of a partial buyout by a group of investors in which Telefónica participated. As such, it is likely that Telecom Italia's future growth in the region will not imply disputes with Telefónica. In previous months Carso/Telmex/América Móvil and Telefónica launched several bids to acquire control of the company through which Telecom Italia participated in Brazil, by far the largest market in the region.

The dispute between the two dominant telephone companies in the region reveals part of the fundamental competition for private investment in BPSs. A part of this investment has been used to control existing markets under oligopoly or poorly regulated conditions. As such, a large part of these investments do not imply a growth in services. The control of these markets includes investment to replace equipment and other necessary maintenance, although coverage is not increased. The development of a new higher profit market, such as cellular telephony, has occurred despite the fact that levels of density and coverage in traditional or principle telephone lines have not reached those of other countries.

UNDP data highlights this fact: in the period from 1990 to 2004 in all of Latin America the ratio of principle telephone lines per 1000 people grew from 61 to 179. In highly developed OECD countries the figures were 462 and 551, respectively, during the same period. The growth of cellular phones follows from these same conditions. In 2004 770 per 1000 people were paying cellular phone clients in these countries, while in Latin America the figure was 319 (UNDP, 2006). There is thus a greater growth in cellular phone services, although the distance from the coverage achieved in more highly developed countries continues to be considerable. In addition, the scarce coverage of principle telephone lines indicates that there are regions with very few telephones, revealing the persistence of strong regional inequalities, which are a sign of the lack of development in Latin American countries.

Brazil and Mexico's situations follow the regional tendency fairly closely. Both countries undertook a complete privatization of telephone services and in both countries companies have incorporated new means for the transmission of data, images and voice. It is currently possible to offer Internet, telephone and television services through existing cable networks. In addition, the dominant telephone company can also transmit Internet and television, although these are very concentrated investments that attend to a reduced segment of the population. Mexico's form of telephone privatization allowed the dominant company to operate on a national level, including long distance and urban telephone services and included a commitment to a minimum level of rural telephone service development. The company was also given licences to exploit the cellular phone market throughout national territory. The result has been a poor advance in the coverage of principle telephone lines, growing from 64 per 1000 inhabitants in 1990, just before the privatization, to 174 in 2004. In 2004 a total of 370 per 1000 inhabitants were paying cellular phone clients. Nonetheless, the company has been investing considerably by acquiring other Latin American firms.

Privatization was undertaken years later, in 1998, and in a much different form in Brazil. The country was divided into various regions in which different companies could provide urban telephone services. In addition, an entirely different company, Embratel, was created for long-distance service. The provision of cellular telephone services was also divided into various regions (Vidal, 2001). There was an ample diversity of buyers, including various telecommunication firms from other large countries (See Table 6.3 and Table 6A in the Appendix). After several years almost all participants in the purchase of privatized companies or in the licensing of cellular phone services had left the market. On the other hand, Telefónica has multiplied its investments, buying various firms that left Brazil as well as acquiring the capital of national actors that left the sector. As mentioned before, Telecom Italia also maintains a presence in the sector through partnerships with local capital. The other player in the market is Carso/Telmex/América Móvil. The results in terms of coverage are very similar to those in Mexico. For every 1000 inhabitants there were 63 principle phone lines in 1990 and 230 in 2004. Paying customers of cellular services totalled 357 in 2004.

In Peru, Telefónica merged the two privatized companies and made various investments to widen cable television, cellular, satellite and Internet services. Outstanding among the investments was the 1998 project to create a pan American fibre optic network from Chile to Panama, including Argentina and Bolivia, which would connect with the United States. The Peruvian subsidiary would operate the cable network and therefore make diverse investments towards this goal, particularly in Peru, even though the investments necessary to widen the network of telephone services in the country did not receive an equivalent priority. In 1990, there were 26 principle telephone lines per 1000 inhabitants. By 2004 this number had grown to 74. In terms of cellular phones, paying customers totalled only 148 per 1000 inhabitants in 2004.

The lack of traditional or principle telephone coverage in the region occurs even though this sector, as highlighted previously, attracts the largest part of resources invested by companies in basic public services. In other areas deficiencies are as important if not more so. In all, while different governments of the region privatize companies, public investment decreases, both in general terms and specifically in the cases of BPSs and infrastructure. In Argentina, Bolivia, Brazil, Chile, Colombia, Costa Rica, Mexico and Peru the average public investment in infrastructure fell from more than 3 per cent of GDP in 1988 to 1.6 per cent in 1998 (Servén, 2005, cited in Fay and Morrison, 2005, p. v). In the following years an significant increase is registered.

However, the greatest problem is that there has not been the relevant increase in investment as billed by the defenders of the Washington Consensus. In 1998, when private investment in infrastructure reached its maximum relative weight, and without distinguishing its objective, it totalled 1.7 per cent of the region's GDP. As has been highlighted in previous pages, a very relevant part of these resources have been used to acquire privatized companies and another part to purchase previously privatized companies. According the figures in Table 6.6, as of 2002 many of the most active companies in BPSs had utilized 50 per cent or more of their total investments in the acquisition of government assets. This is the case of the electric energy sector companies such as AES, EDF, Iberdrola and Endesa. Other firms that dedicated a larger proportion of their resources towards new investments have abandoned the region. As such, after 1998 the ratio of private investment in infrastructure to GDP is even lower. Currently, investment with the private participation in infrastructure (PPI) does not surpass 0.9 per cent of the region's GDP and is earmarked, in its majority, toward the energy and telecommunications sectors. 'Of the 0.9 per cent of GDP, 43 per cent was earmarked for the energy sector, 41 per cent for telecommunications, 5 per cent for transportation and 1 per cent to water and water treatment' (Fay and Morrison, 2005, p. viii).

Taken together, public and private investment in infrastructure barely reaches 3 per cent in the region, a figure well below that of the 1980s. Globally, the result is that investments made in infrastructure are insufficient to meet the needs of the region's growth. A recent study published by the WB highlights the need to increase investments in infrastructure two-fold for the region merely to prevent further losses in competitiveness to regions such as East Asia and the Pacific. According to the study, for Latin America and the Caribbean to reach the level of coverage of South Korea over a period of 20 years or simply to achieve the level that China maintains, the required percentage is between 4 and 6 per cent of GDP (Fay and Morrison, 2005). In 1980 the coverage of productive infrastructure – roads, electricity and telecommunications, was greater in Latin American and the Caribbean than in South Korea and other countries in Southeast Asia. Today not only is the region's coverage much smaller than that of these countries, but also than that of middle-income countries and China (Fay and Morrison, 2005).

Upon considering the unmet necessities in previous years and the economic growth of various developing countries, it is clear that the amount of resources devoted to investment in infrastructure and public services should be greater than those estimated in WB studies. In a scenario of economic growth of more than 6 or 7 per cent of GDP, investment in infrastructure and public services of 8 per cent is a necessary benchmark. Currently, in as much as the South American economies that are distancing themselves from Washington Consensus proposals register greater growth, increases in public investment appear as an undeniable fact. In this context several measures have been taken to re-establish diverse public companies in telecommunications, energy and petroleum. Over ten years of experience in systematic privatizations reveals that they do not produce a sufficient increase in private investment to offset foregone public investment. To meet the necessities arising from a greater economic growth in the region it is necessary that investment in infrastructure equal at least 8 per cent of GDP.

Capital formation, public finances and privatizations

In 1980 fixed investment was 27.6 per cent of GDP. In the following years there was a tendency towards a diminishing gross formation of capital as a percentage of GDP, registering around 20 per cent that has not shown significant modifications to date. At the beginning of the 1980s the largest economy in the region had a coefficient of investment of 32.2 per cent. At the beginning of the 1990s, when the first push towards privatization was seen, the capital to GDP ratio was around 20 per cent. In the following years the numbers did not vary much.

The national privatization plan in Brazil had a strong impact in 1997 and 1998. These two years account for 62 per cent of the programme's total revenues (BNDES, 2005). As can be seen in Table 6.3 and in the Appendix, in addition to the telephone companies, several electrical power generation companies were also privatized. If the privatizations were a positive element to foster investment and given that there were no restrictions places on external capital entry there should have been an increase in the coefficient of investment or at least its decline should have been avoided. However, this was not the case. According to ECLAC's data (2007), in 1998 the gross formation of capital as a proportion of GDP was 18.2 per cent, which diminished to 16.7 per cent in 1999. In 2002 it had shrunk to 15.4 per cent and in 2003 was even lower, at 14.5 per cent. It is only in the following years, after the Brazilian government suspended its national privatization programme and adopted policies that differed slightly from those proposed by the Washington Consensus, that a small increase in the coefficient of investment appeared.

Brazil's case is not an exception in the region. Mexico's gross formation of capital as a proportion of GDP is approximately 20 per cent, despite far reaching modifications in legal codes allowing the unrestricted entry of foreign capital and the maintenance of a consistently pro-privatization stance by the country's successive governments. During Fox's presidency, from 2001 to 2006, total public investment equalled 2.5 per cent of GDP. However, this figure includes investments made by the private sector, not only in the construction of electrical power generation plants, which are offered through auctions by the state company Comisión Federal de Electricidad (CFE), but also in petroleum exploration projects offered through PEMEX. In Mexico publicly financed projects can be considered under the public budget as Long Term Productive Infrastructure Projects with Deferred Impact on Registered Expenditures (PIDIREGAS is its Spanish acronym). According to information by the CFE (2007), at the end of 2006 the present net value of all projects financed through PIDIREGAS was 43 billion dollars, measured in January 2007 exchange rates. PEMEX reported an external debt through PIDIREGAS of 38 billion dollars at the end of 2006 (Banco de México, 2007). The public external debt as of December 2006 was 42 billion dollars.

The payment of PIDIREGAS debt is booked to the public budget and has represented around 1 per cent of GDP in recent years. Within the budget's public sector expenditure, investments appear under two distinct categories: a) budgeted physical investment and b) fostered investment (SHCP, 2006). The first is carried out with public sector resources, and includes federal government investment, physical investment in communications and transport, PEMEX's investment fund and other investments. During the period of 2000 to 2006 this category did not grow, totalling in the last year less than 2 per cent of GDP. The so-called fostered investment through the public sector is defined as 'the total of budgeted physical investment and that which is made through off budget expenditures, the net of PIDIREGAS amortizations...' (SHCP, 2006, p. 191). This implies that investments made by private companies are also considered and includes a diverse range of financial commitments for the public administration. The Federal government estimates that for the period of 2007 to 2012, excluding the possibility of substantial changes, budgeted physical investment will not increase and that the estimated increase to 5.2 per cent of GDP will be reached through private sector investment with the respective growth in public debt. As such, there is an almost complete withdrawal of public investment, yet without a corresponding increase in private investment. Investment relative to GDP does not register a significant increase, even in estimates for the coming years, which suppose a maintained economic growth of approximately 3.5 per cent of GDP.

Argentina experienced a reduction of even greater proportions. By 1994, after the bulk of privatizations had taken place and even after many of YPF's extraction zones had been sold in parts along with the company's capital in the stock market, the coefficient of investment was 24.2 per cent. In the following years there was a systemic fall. In 1998, according to ECLAC's information (2007), the figure had fallen to 19.1 per cent. In 2000 it had been reduced to 16.2 per cent and by 2002 it stood at 10.2 per cent. In this same year public investment equalled 0.7 per cent of GDP. This reduction was systematic during the entire decade of the 1990s. The continuity of the country's economic policy, which included widespread privatization, resulted in a drastic reduction of capital formation as a percentage of GDP in the midst of a sharp crisis that resulted in a group of foreign companies operating in public services declaring a moratorium on their payments.

As ECLAC (2003, p. 36) highlights, the policy of indiscriminately opening the Argentinean economy during the 1990s and until 2002, which was accompanied by strong incentives for foreign companies to obtain attractive profits, was unsustainable in the long term. The firms that declared a payment moratorium in 2002 or that closed their subsidiaries belonged to the sectors that remitted the largest amounts of profits during the 1990s: public services and petroleum. From 1992 to 2001 profits sent abroad by foreign companies operating in Argentina totalled over 19 billion dollars, equal to 24.7 per cent of net FDI inflows during the same period. Petroleum, non-financial services and banks accounted for 75.3 per cent of profits transferred abroad.

The establishment of incentives for companies, including foreign ones, to invest is part of the economic policy executed in other Latin American countries. The Programa Prioritario de Termoelectricidad (PPT) was established in Brazil in 2000 with various incentives for private investment. However, given the poor response by private enterprise, an inter-ministerial agreement to increase the incentives of the PPT was established, including a guaranteed fixed price for natural gas and the granting of an exchange rate guarantee for projects in the scheme that utilized Bolivian natural gas. The goal was for the private sector to construct 49 electrical power generation plants, of which 44 would be gas fed (*Oil and Gas Journal Latinoamerica*, 2002). These incentives met with partial success in the midst of a process of asset sales by transnational companies in the electricity sector. The decision to reduce participation in Brazilian companies was a result of a fall in profits due to the Real's devaluation and financial problems that several companies faced in their home countries. Among these companies were Enron, AES Corporation, Eléctricité de France (EDF) and Pennsylvania Power and Light (PPL). It is noteworthy – as ECLAC highlights (2003, p. 47) – that among possible buyers and actual buyers there were no transnational companies. Much to the contrary, it was Petrobras that increased its investments. As was mentioned in before, the Brazilian firm acquired shares of Pérez Companc in Argentina, while in Brazil it bought Enron's shares. Although the government sold a part of the company's shares on the stock market, it maintains a majority in the board of directors. Currently Petrobras is actively internationalizing in the region, confirming the recent advance of companies fully or partially owned by the state in the region's energy sector. In what follows, other examples will also suggest the strengthening of various state-owned companies through investment in infrastructure.

However, it is first necessary to mention that the pullback in investments in infrastructure and public services and its concentration in activities with significant incentives or far-reaching government guarantees is also present in other countries. In Colombia in 2002 strong differences were present between several transnational telecommunications companies and state-owned Telecom Colombia. The expected income from the operation contracts signed in 1993 had not been obtained, resulting in legal actions demanding a compensation of 1.5 billion dollars. In the following years the organizational scheme of Colombian telecommunications changed and the firm Telecom Colombia passed into the hands of Telefónica, as had happened in other countries. This firm now owns 50 per cent of the company's stock, while the rest remains in the hands of the Colombian state. Among the operational details it is noteworthy that Telefónica will receive from Telecom Colombia a fixed quantity of 7 per cent of the EBITDA (Earnings Before Interest, Taxes, Depreciation and Amortization) between 2006 and 2010 and of 3 per cent from 20011 to 2022 in the concept of a managerial bonus, under the condition that the EBITDA exceeds 115 per cent of the consideration (Telefónica, 2006). The acquisition is also accompanied by other terms that guarantee greater profits for the foreign firm.

The electricity sector presents a similar case. The privatization of the firm Empresa Energía de Bogota (EEB) was partial and required the modification of various laws to permit a distinction between various activities in the sector, among them the generation, transmission and commercialization of electricity. The majority partner of EEB is the city of Bogota, even though private capital also participates. Among the various subsidiaries of the company is a mix of state and private entities and capital, including foreign partners. The firm in charge of electricity generation is distinct from that which commercializes and distributes electricity. EEB focuses on the transmission of electricity, even while other companies with state participation perform the same activity. To summarize, there is a diverse and complex restructuring of the electricity sector that without a doubt opens possibilities for private investments, some of which carry specific guarantees and incentives.

In Mexico, there is also a set of modifications in place that allows private investment in electric power in a context in which the state companies shoulder risks. Through modifications in laws and secondary regulations, companies defined as private are allowed to construct electrical power generation plants fed by natural gas. All energy produced is sold by the two state companies, Comisión Federal de Electricidad (CFE) and Luz y Fuerza del Centro (LFC). The power generated by these producers is bought by CFE and LFC independently of the levels of consumption in the country. In addition, the pricing calculation for the purchase of electricity includes the costs of energy inputs and all variable operating and maintenance costs incurred in the generation and transmission of energy up to the connection point with CFE's network. Also considered are costs relating to water, chemical products, lubricants, and handling of ashes and combustible material. Increases in any of these costs are immediately passed on to the prices of electrical power purchased by the state companies. In terms of natural gas there is a full supply guarantee by state-run PEMEX, even though its production is insufficient in the country as a whole. Although Mexico currently imports natural gas, its higher prices do not translate into any loss in profits for the firms that participate in electric power production in Mexico. Among the companies that have invested in this sector are Iberdrola, Unión Fenosa, Mitsubishi, Siemens and Electricité de France.

Other legal modifications have and incentives have been established to foster the growth of private investment in the region. It is therefore possible to sustain the argument that privatizations have been part of a broader set of measures used to reach this goal. In this sense, the multiplication of bilateral investment agreements between governments of the region and the Unites States and several European countries, as well as free trade agreements with broad chapters regarding investments, are particularly noteworthy. Between 1990 and 1999, when the vast majority of privatizations in the region took place, 55 bilateral investment treaties (BIT) were signed, along with six free trade agreements, among them CARICOM, The Group of Three, MERCOSUR, NAFTA and CAFTA, and six bilateral free trade agreements between Latin American countries. The majority of BITs were between a Latin American country and the Unites States or a European country. As such the substantive theme is the guarantees offered for foreign direct investment in Latin America. In the following years other BITs have been signed, principally between the United States and countries with which previous agreements had not existed.

An analysis of BITs is obviously broad and diverse. For the purposes of this text, their most noteworthy elements are the guarantees and assurances established for foreign investment. Among them are (a) most favoured nation treatment as a precondition to their establishment, which is maintained once established; (b) the absence of performance-based conditions within the countries that sign the commercial agreement; (c) the establishment in the majority of the treaties that the most favourable legislation for investors will prevail even when this factor is not explicit in the treaty; (d) differences between signing parties in terms of the application of the treaty will be submitted to ad hoc tribunals, forgoing the application of national laws; (e) no limitations on the transferrals of funds within the country and to other countries by the investors of a signing party to the treaty, including income in its broadest concept, payment of contracts and revenues produced from total or partial liquidations of contracts; (f) transferrals should be executed in a freely used currency and no restrictions are envisioned in the case of balance of payment difficulties, with the exception of a few treaties, but only temporarily in an equitable and non-discriminatory fashion (ALCA, 1999). These are guarantees that the United States Department of State establishes, supervised by an office established as part of the organization to attend to investments by its co-nationals abroad. (US Department of State, 2005).

In the months preceding the Argentinean crisis, and during its development, foreign investment in the country benefited from these agreements. In disputes between the Bolivian government and Spanish transnational companies these clauses have been applied as well. These are part of the set of new conditions created during the 1090s as a result of the execution of the Washington Consensus agenda. They are measures, much like privatizations, that sought to encourage private investment, which would be the basis for greater growth in the economies of the region. However, as has been shown here, this has not happened.

The lack of dynamism in investment has been seen as an important problem given the growth in the region in the last three years, and responses have obviously not been homogeneous. But there are also other processes resulting from privatizations and the opening to foreign capital with specific incentives that are influencing the behaviour of Latin American and Caribbean economies and that can negatively affect their performance in the coming years. In the period of 1991 to 2000 remittances sent abroad in the form of profits from foreign companies equalled 0.6 per cent of GDP in the group of South American countries. Starting in 2001 there has been a rise in this indicator. In 2006 profits represented 2.7 per cent of the product. In Mexico and Central America the figures are lower, but they also showed an important increase, rising from 0.5 per cent to 0.9 per cent in the respective periods (Machinea, 2007).

On the other hand public spending has been conducted with strict adherence to IMF and WB policies, maintaining a primary budget surplus. In 2005, based on information from ECLAC (2007), for all of Latin America and the Caribbean, the primary surplus of the public sector was 3.1 per cent of GDP. This is a figure slightly above that of the region's different central (excluding state and provincial) governments' interest payments as a proportion of GDP. The overall result for public sector balances is a small surplus. The same result was produced in 2006, with the exception of Brazil, which registered a deficit of 3 per cent of GDP. However, this deficit was not the result of an increase in spending or public investment. Rather, it was a result of interest paid by the Brazilian government. In 2006 the interest paid by the central government of Brazil equalled 5.3 of GDP. The country's economy is operating with very high interest rates. In 2006 the market representative interest rate paid on assets was on average 40 per cent with an annual variation of 3.1 in the consumer price index.

In Brazil, as in the majority of the countries of the region, internal credit to the private sector is not a relevant factor in economic growth. With varying importance, bank privatization has taken place in much of the region. The most important case is Mexico, as is highlighted in the second part of this chapter. After the privatizations undertaken at the beginning of the 1990s, almost all commercial banks would subsequently be sold to foreign capital. Five foreign banks, Citibank, BBVA, Santander, Scotiabank and HSBC, currently hold more than 80 per cent of commercial bank assets. Foreign capital is also prevalent in insurance and pension funds, the latter also privatized. This has not meant a growth of internal credit towards companies. Internal credit issued to companies in 2000 on average equalled 16.7 of GDP; in 2006 it was 16.4 per cent.

In Brazil internal credit to companies in 2006 was 38.1 per cent, while in 2000 it was 30.6 per cent. This is a small increase but under conditions of very high interest rates. However, it is important to observe the strengthening of several domestically-owned banking groups. Alongside this strengthening is that of various business groups, as mentioned in the third part of this chapter. Several of these groups are the result of privatizations and currently play an important part in the relationship between the Brazilian and world economy. One of them, Embraer, is an important world-wide competitor in the manufacture of medium-sized passenger aeroplanes. This group of businesses is without doubt a result specifically related to the privatizations of the country. As of today their growth has not been accompanied by a higher coefficient of investment, with a relevant growth in national currency denominated credit or an important advance in infrastructure investment. Nonetheless, changes could occur in the fairly immediate future.

The slight increase in the coefficient of investment registered in the region's balances starting in 2005 has resulted principally from the gains made in several countries that are characterized as having applied an economic policy that diverges from the proposals of the Washington Consensus and in those that have suspended privatizations. These are countries that are promoting public investment and that have applied various mechanisms to recuperate activities that had been sold or offered by concession to private companies. It is within this context that the Argentinean government created Energía Argentina S.A. (ENARSA). On the other hand the Bolivian government has strengthened Yacimientos Petrolíferos Fiscales de Bolivia, recuperating the revenue from natural gas and establishing new contracts with firms that had previously extracted it. In Venezuela the government acquired CANTV, Electricidad de Caracas and PDVSA recovered the Orinoco Belt. The argument maintained by various governments is that petroleum, natural gas, electricity and telecommunications are strategic activities for the growth of their economies. As such they have proceeded to establish state companies in these sectors. Many consider that this proposal must also include potable water and drainage services. In this context, the issue of growth in public investment gains an added importance.

Taken together, it can be seen that despite the insistence from the proponents of the Washington Consensus agenda that it is necessary to continue privatizing and generating new and greater incentives for private investment in infrastructure, the events in several countries are proceeding in a different direction. Even in Brazil, as was earlier highlighted, privatizations have given way to a new mixed formula such as that of Petrobras. Among the region's biggest economies and several medium-sized ones, Mexico is the exception. Privatization and the creation of multiple incentives and preferential treatment for private investment in infrastructure are still insisted upon, as in the case of PIDIREGAS and the external producers of electric power. However, private investment is not growing and the financial burden of these investments on public finances is increasing.

The weakness of capital formation and the limited contribution from public expenditures towards the development of infrastructure in a large number of Latin American and Caribbean countries reveal that privatization has not achieved the results proposed by those who promoted them. The mere fact that a large part of privatized companies pertain to basic public services (BPSs) constitutes a grave problem. In the region discussion over BPSs continues while these services still have very limited coverage. In many countries services such as potable water, drainage and electric power continue to be notoriously insufficient. The social inequality and the weakness of formal job creation underline the fact that these services represent minimum basic needs of the population, which is not compatible with the high levels of profits that some businesses have been able to make though their investments in these sectors.

Summary and conclusions

The execution of economic reforms as recommended by the Washington Consensus has not produced significant economic growth. In recent years several of the economies that are registering robust growth have distanced themselves from Washington Consensus proposals, and have changed government policies in order to resolve problems relating to deficient infrastructure. Mexico is one of the few countries that continue to abide by Washington Consensus economic policy without objections; among the six or seven largest economies of the region, it is also the one that has witnessed the least economic growth.

During the 1990s privatization programmes were undertaken throughout the region. As mentioned earlier, while the sale of state companies took place in different moments with interruptions in some countries, in several areas of activity state companies were acquired by large national companies. In Mexico, and under different rules in Brazil, the intention at the beginning of the privatization process was to foster the growth of national financial and business groups. However, conditions changed during the second half of the 1990s and the entry of transnational firms became intensified through acquisitions of assets in the process of privatization or through the purchase of recently privatized firms.

As the result of both processes, some transnational firms achieved a dominant position in the region. It is noteworthy that many of these firms, both American and European, had only begun internationalizing a few years beforehand. These firms operate in infrastructure and participate in the buying and selling of assets both in their home countries and in other regions. As such, the privatizations in Latin America have been part of a process of cross-border sales and fusions that have had among their results the consolidation of companies that reduce payrolls and whose investments have high import contents. In addition, many of them have been active in new fusions, becoming an important element in the economic concentration that many societies now face. On the other hand, for the economies of Latin America, privatization revenue has not been relevant in terms of GDP or public spending. Moreover, the amount of income obtained through the sale of assets, increasingly to transnational firms, is exceeded several times over by the net outflows of capital from the region in the form of payments of interest and earnings.

As this chapter analyses, in recent years there has been a significant reshuffling of ownership of assets and several of the firms that had invested in the region during the second half of the 1990s and the beginning of the current decade have sold their capital and have withdrawn from the region. One of the results of this evolution has been the increased dominance of several firms in the region. Notable cases in this regard include the telecommunications sector, with two companies dominating investment throughout the continent, and the electricity and water services sectors.

The companies involved have only recently become internationalized, and the European ones until quite recently were state-owned companies. Besides this, large capital firms in many Latin American countries seem to affirm their position of partners or rentiers, and even in cases where they have become strengthened, such as in Brazil, they have not undertaken the investment in infrastructure and basic public services that would allow the region to maintain conditions of international competitiveness. Even in telephony investment has been part of a process of economic concentration that currently has two dominant players: Carso-Telmex-América Móvil, controlled by Mexican capital, and Telefónica, of Spanish capital. Taken together, privatizations and the realization of an economic policy that promotes private investment in infrastructure have not translated into increases in the coefficient of investment. The ratio of gross capital formation to GDP has not increased and changes are only evident in countries that are not carrying out Washington Consensus policies and that have suspended privatizations or have even acquired previously sold assets or restored the rules so that public investment is maintained in public services and in energy companies.

Infrastructure and basic public services continue to present significant deficiencies. It has become clear that investments in this sector have been notably insufficient and the conditions that have been created to assure the participation of private capital have guaranteed earnings for the companies, but not necessarily an enlargement of services. Such are the cases of investment in several telecommunication and energy generation firms. In order to reach sustained economic growth such as that achieved by South and Pacific Asian countries it is necessary to substantially increase the resources dedicated to infrastructure. Even the WB has warned of this necessity. The experience of recent years in Latin America indicates that private investment has not been sufficient and has not necessarily extended the coverage of these activities. The decision of several governments of the region is casting new light on the positive role of public investment in these areas.

Appendix

Table 6A Latin America largest privatizations with foreign investors, 1987–2005

Privatized firm	Country	Year	Acquiring firm	Amount of acquisition (1)
YPF S.A.	Argentina	1999	Repsol, Spain (84%) acquires stocks in the market and those held by other shareholders.	13.2
Aeropuertos Argentinos	Argentina	1998	Empresas USA	5.1
Telesp Participacoes S.A.	Brazil	1998	Telebrasil Sul Participacoes S.A. Consortium: Telefonica S.A. and Subsidiaries (56.6%), Iberdrola Investimentos (7%), BBV(7%), RBS Participacoes Telecom de Portugal (23%) (6.4%); Spain (70.6%); Portugal (23%), Brazil (6.4%)	5
Petrobras	Brazil	2000	Parcial sale. Included ADRs and equities in Sao Paulo and Madrid	4.2
Banco do Estado de São Paulo (BANESPA)	Brazil	2000	Banco Santander Central Hispano, Spain	3.55
Telefónica del Perú	Peru	2000	Telefónica, Spain (56.7%)	3.218
Telesp Celular	Brazil	1998	Telecom Portugal (100%)	3.1
Telmex	Mexico	1991	Southwestern Bell/France Cable et Radio/Grupo Carso	2.8
Light Servicos de Electricidade	Brazil	1996	AES, USA (20.3%); Houston, USA (20.3%), EDF, France (20.3%); BNDES, Brazil (16.4%)	2.3
Embratel	Brazil	1998	MC I(100%)	2.3

Empresa de Energia de Bogota (EEB)	Colombia	1997	Two groups dominated by Endesa of Spain.	2.18
YPF S.A.	Argentina	1999	Repsol, Spain (14.9%) Repsol acquires stocks remaining in government hands	2
Entel Perú	Perú	1994	Telefonica, Spain (100%)	2
Telefónica de Argentina	Argentina	1990	Telefonica, Spain (10%); Citibank, USA (20%) Others foreign banks (37%)	2
Telecom Argentina	Argentina	1990	Telecom Italy (32.5); France Telecom (32.5%)	1.8
Telecentro Sul	Brazil	1998	Techold Part, Italy (19%)	1.8
Compania Anonima Venezuela (CANTV) Telefonos de	Venezuela	1991	GTE/AT&T Consort./Telefonica de Espana (72%)/ local investors (28%)	1.8
COELBA	Brazil	1997	Iberdrola, Spain (39%)	1.8
Elektro	Brazil	1998	Enron, USA (100%)	1.7
Electricidad de Caracas	Venezuela	2000	AES Corp. (81.3%)	1.658
Telmex	Mexico	1990	Carso, Mexico, Southwerstern Bell, USA and France Telecom (20.4% 'AA')	1.7
Cia de Electricidade do Estado da Bahia	Brazil	1997	Iberdrola (39%)	1.6
Reserva de Gas Camisea	Peru	2000	Pluspetrol Energy, Argentine; Hunt Oil Co, USA and SK Co, Korea	1.6
Entel/CPT	Peru	1994	Consortium led by Telefonica de Espana (Spain) (100%)	1.4
Companhia Centro-Oeste de Distribuicao de Energia Eletrica (CCO)	Brazil	1997	AES Corp.	1.4

(*Continued*)

Table 6A (Continued)

Privatized firm	Country	Year	Acquiring firm	Amount of acquisition (1)
Ferrocarril del Noreste S.A. de C.V.	Mexico	1997	Transportacion Ferroviaria ulMexicana S de R.L. de C.V. (MEX) and Kansas City Southern Industries, Inc. (USA)	1.4
COELCE	Brazil	1998	Endesa, Spain (37.5%); CERJ, Brazil (36.5%)	1.3
Aerolíneas Argentinas	Argentina	1990	Iberia Airlines, Spain (69.4%)	1.3
Elektro Eletricidade e Servicos S.A. - Elektro	Brazil	1998	Enron International; USA (100%)	1.27
Ferrocarril Pacifico Norte (Hoy Ferromex)	Mexico	1997	Union Pacific USA (26%)	1.2
Empresa Bandeirante de Energia	Brazil	1998	CPFL-Brazil (Votorantin, Bradesco e Camargo Correa) (44%), Electricidade de Portugal (56%)	1.2
Sidor (CVG Siderurgical del Orinoco)	Venezuela	1997	Amazonia Consortium led by Hylsamex of Mexico (30%) and Sivensa of Venezuela (20%). Â Consortium also contains Tamsa of Mexico (17.5%), Usimas of Brazil (10%), Argentinian Siderar (17.5%), and Argentinian Techint (5%).	1.2
Telesudeste Celular	Brazil	1998	Telefonica	1.2
Banco do Estado de São Paulo (BANESPA)	Brazil	2001	Banco Santander Central Hispano, Spain (completo 100%)	1.1

Companhia de Geracào de Energia Elétrica de Paranapanema	Brazil	1999	Duke Energy Corp, USA (100%)	1.1
Comgas	Brazil	1999	British Gas (70%), Shell (26%), England	1.1
Compania Anonima Telefonos de Venezuela (CANTV)	Venezuela	1996	Public offer. Foreign investors (79%)	1.143
Cemig(Minas Gerais	Brazil	1997	Southern Electric USA	1.1
Companhia Energética de Pernambuco (CELPE)	Brazil	2000	Iberdrola, Spain (79%)	1.1
Cia Riograndense de Telecomun	Brazil	1998	Telefonica do Brasil Holding: Telefonica de Espanha and affiliates CTC, Tasa e Citicorp (83%), Rede Brasil Sul Participacoes – RBS (17%); Brazil (17%), Spain (83%)	1.1
Aseguradora Hidalgo	Mexico	2002	Metropolitan Life Insurance, USA (100%)	1
Companhia de Eletricidade do Estado de Rio de Janeiro	Brazil	1996	Eletricidade de Portugal (30%), Electricidad de Panama (30%)	1

Note: (1) billion dollars.

Source: Own elaboration supported by UNCTAD, World Investment Report 2000, United Nations, Ginebra, 2001, p. 134; BNDES, Privatizações no Brasil 1990–2004, Rio de Janeiro, February, 2005; ECLAC, *La inversión extranjera en América Latina y el Caribe*, 2002.

References

ALCA (1999) *Acuerdos sobre inversión en el hemisferio occidental: Un compendio,* Unidad de Comercio de la OEA, en http://www.ftaa-alca.org/ngroups/ngin/publications/spanish99/tr_lists2.asp

Andres, L., Foster, V. and Guasch, J.L. (2005) *The Impact of Privatization in Firms in the Infrastructure Sector in Latin American Countries',* World Bank, Washington.

Banco de México (2007) *Informe Anual 2006,* Banco de México, México.

BNDES (2005) *Privatizaçoês no Brasil 1990 / 2004,* BNDES, Rio de Janeiro.

Cardoso, Eliana (1992) 'La privatización en América Latina', in *¿Adonde va América Latina?,* Vial, Joaquín (ed.), Corporación de Investigaciones Económicas para Latinoamérica Editor, Santiago, Chile.

CFE (2007) *PIDIREGAS. Informe al cuarto trimestre de 2006,* Comisión Federal de electricidad, México.

Chong, Florencio and López de Silanes, Alberto (2003) 'The Truth about Privatization in Latin America', Research Network Working Paper #R-486, Inter-American Development Bank and Latin American Research Network, Washington.

De Selys, Gérard (1995) 'Privé de public', Editions EPO, Bruxelles.

ECLAC (1998) *Balance preliminar de las economías de América Latina y el Caribe 1998,* United Nations, Santiago de Chile.

ECLAC (2002) *La inversión extranjera en América Latina y el Caribe, 2001,* United Nations, Santiago de Chile.

ECLAC (2003) *La inversión extranjera en América Latina y el Caribe, 2002,* United Nations, Santiago de Chile.

ECLAC (2007) *Estudio Económico de América Latina y el Caribe, 2006–2007,* United Nations, Santiago de Chile.

Fay, Marianne and Morrison, Mary (2005) 'Infrastructure in Latin America & the Caribbean: Recent Developments and Key Challenges', Finance, private sector and infrastructure unit, World Bank, Washington.

Galal, Ahmed and Shirley, Mary (1994) *Does Privatization Deliver? Highlights from a World Bank Conference,* EDI Development Studies, World Bank, Washington.

Gamir, Luis (1999) *Las privatizaciones en España,* Ediciones Pirámide, Madrid.

IADB (1996) *Informe 1996,* Inter-American Development Bank, November, Washington.

IADB (1998) *Informe 1998,* Inter-American Development Bank, November, Washington.

Krueger Anne (1990a) 'Government Failures in Development', *Journal of Economic Perspectives,* 4(3), Summer.

Krueger Anne (1990b) 'Theory and Practice of Commercial Policy: 1945–1990', National Bureau of Economic Research, Working Paper 359.

La Porta, Rafael and Florencio López-de-Silanes (1999) 'The Benefits of Privatization: Evidence from Mexico', *Quarterly Journal of Economics,* 4: 1193–242.

LatinTrade (2003) 'Informe regional. Las 100 mayores fusiones y adquisiciones' in *Latin Trade,* 11(3), Coral Gables, FL.

Lieberman, Ira and Fergusson, Robert (1998) 'Overview of privatization and emerging equity markets', in *Privatization and Emerging Equity Markets,* Lieberman, Ira and Kirkness, Christopher (eds), World Bank and Flemings, Washington.

Lora, Eduardo (2001) Las reformas estructurales en América Latina: Qué se ha reformado y cómo medirlo, Research Department Working papers 462, Inter-American Development Bank, Washington.

Machinea, José Luis (2007) *Estudio Económico de América Latina y el Caribe 2006-2007.Presentación del Secretario Ejecutivo*, Comisión Económica para América Latina, United Nations, Santiago de Chile.

Minsky, Hyman (1986) 'The crisis of 1983 and the prospects for advanced capitalist economies', in *Marx, Schumpeter and Keynes: A Centenary Celebration of Dissent*, Kelburn, Suzanne and Bramhall, David (eds), M.E. Sharpe Inc. New York.

OECD (2002) 'Recent privatization trends in OECD countries', *Financial Market Trends*, num. 82, June, París.

Oil & Gas Journal Latinoamérica (2003), 9(5), September.

Rischard, Jean-François and Garrett, William (1998) 'Foreword', in *Privatization and Emerging Equity Markets*, Lieberman, Ira and Kirkness, Christopher (eds), World Bank and Flemings, Washington.

SHCP (2006) *Criterios generales de política económica 2007*, Secretaría de Hacienda y Crédito Público, México.

Shleifer, Andrei, and Vishny, Robert (1996) 'A survey of corporate governance', NBER Working Paper 5554. Cambridge, MA: National Bureau of Economic Research.

Telefónica (2006) Press release, Telefónica adquiere el control de Colombia Telecom, http://www.telefonica.com.co/compania/ 7 April.

UNDP (2006) *Report on Human Development*, New York: United Nations Development Programme.

US Department of State (2005) *U. Bilateral Investment Treaty Programme*, at: http://www.state.gov/e/eb/rls/fs/22422.htm

Vickers, John and Yarrow, George (1988) *Privatization: An Economic Analysis*, Cambridge, MA: MIT Press.

Vidal, Gregorio (2001) *Privatizaciones, fusiones y adquisiciones las grandes empresas en América Latina*, Anthropos Editorial, Barcelona.

Williamson, John (ed.) (1990) *Latin American Adjustment. How Much Has Happened?* Institute for International Economics, Washington.

World Bank (1997) *El Estado en un mundo en transformación, Informe sobre el desarrollo mundial 1997*, Washington.

World Bank (2000) 'Progress in privatization, 2000' in *Global Development Finance 2000*, Washington.

World Bank (2001) 'Progress in privatization in developing countries', in *Global Development Finance 2001*, Washington.

Index

accounting
    accountancy industry 25–6
    off balance sheet 49, 50–1
    Private Finance Initiative (PFI) 48–52
Accounting Standards Board 52
Arrow–Lind theorem 55

banks, project failure 16, 22
Boateng, Paul 46–7
borrowing, public *see* public borrowing
Brown, Gordon 10
budget deficits 46, 48, 50
Build-Operate-Transfer (BOT), water industry 99–100

capital employed, profit 28
capital expenditure, public investment 46, 50
capital values 5, 21, 44
capitalism, globalization 29–30, 32
commercial confidentiality 11, 25, 26
competition
    competitive pressure 4, 6
    ex ante 6
concentration
    construction industry 6
    electricity industry 175–8
    Latin America 210–15
conflict of interest 9–10, 25, 26
contracts
    contract drift 19
    inflexibility 61–2
    leases 44
    negotiation 69
    PFI issues 67–71
    principal–agent 68
    specification 68–9
    sub-contracts 20, 21
    transaction/monitoring costs 68–9
    uncertainties 17
    water industry 85
conventional public investment
    front loading 50–1
    PFI distinguished 40–1
    risk 54–5
    see also public investment
corporate taxation, welfare state 28–9
costs
    additional costs 19, 22
    additionality 23–4
    affordability 10–11, 20
    alleged savings 12
    annual payments 21, 23, 44–5, 49
    bidding 6, 15
    capital values 5, 21, 44
    discount rate 7, 8, 45, 51
    distributional conflict 24
    estimates 11
    financial costs 7, 8, 18–24, 52–4
    net present cost (NPC) 7, 8
    overruns 10, 14–15, 17, 55
    private/public debt 3, 54–5
    professional fees 3
    refinancing 16, 19, 22, 53
    risk premium 7–8, 11, 20, 21, 54
    sunk costs 62
    uncertainty 4, 11
    whether lower under PFI 63–7
    see also value for money

Denmark, electricity industry 128–9, 131–2, 136
design, build, finance and operate (DBFO) 3, 13, 20–1, 24
developing countries, water industry 79–80, 84, 115
Doronzo, Raffaele 121–59

economy
    economic management 204–10
    national regulation 27, 30

electricity industry
  competition reports 168–9
  concentration 175–8
  consumers 121–59
  consumption 162, 164, 165
  Denmark 128–9, 131–2, 136
  deregulation 160–201
  directives 127–8, 167–9
  dynamics 140–4
  EU-15 reforms 127–37
  EU-25 structure 162–5
  European Union (EU) 121–59
  external trade 129, 162, 165
  findings 149–57, 192–3
  Finland 128, 129, 136
  France 128, 129–30, 133–4, 190–2
  Germany 128, 129, 131, 134–5, 187–9
  Italy 128, 129, 130–1, 134, 189–90
  large countries 180–92
  liberalization 124, 129–30, 132–3, 167–80
  liberalization process 169–80
  market capitalization 181
  market design 169–74
  market design indicators 170–1
  market structure 132–7
  national champions 161
  national laws 132
  objective/subjective evidence 126
  over the counter markets (OTC) 173
  privatization 122–6, 160–201
  production 162, 163, 165
  quality 150
  reform agenda 123–4
  reform paradigm 121–59
  reform trends 129–32
  regulation 127–8, 167–9
  regulatory indices 139
  restructuring 180–92
  retail markets 176
  Spain 128, 129, 130, 134, 184–7
  structural market indicators 176
  structure 162–5, 180–92
  supply sources 128–9
  Sweden 128, 129, 135, 136
  traded volumes 174
  transmission network 174–5
  unbundling 122–7, 130–3, 135, 152, 154–7
  United Kingdom 128, 129, 132–3, 182–4
  wholesale markets 176
  World Bank 123–4
electricity prices
  analysis data 137–9
  average partial effect 148
  consumer price index (CPI) 147
  consumer satisfaction 144–57
  descriptive statistics 145
  panel estimation 126, 141–2
  price evolution 178–80
European Union (EU)
  electricity directives 127–8, 167–9
  electricity industry 121–59
  internal market 125, 167, 168
  Stability and Growth Pact 48–9
  water industry 77, 104–6, 109

financial advisers
  accountancy industry 25–6
  Big Four 25, 26
  civil servants 25
  private finance 1, 3, 6, 10
Finland, electricity industry 128, 129, 136
Fiorio, Carlo V. 121–59
Florio, Massimo 121–59
France, water industry 77–9, 81, 82, 107–8
front loading 50–1, 65

Germany, electricity industry 128, 129, 131, 134–5, 187–9
globalization
  global capitalism 29–30, 32
  information technology 27
  neo-liberalism 27
  transnational corporations (TNCs) 29

Hall, David 75–120
Heald, D. 52, 53, 62
Highways Agency 20–1, 24

information technology
  globalization 27
  PPP projects 27
insolvencies
  Public Private Partnerships (PPP) 10, 16, 19
  vulnerability 55
interest
  future payments 48
  low rates 71
  public borrowing 52, 54, 55, 56
International Monetary Fund (IMF) 204
investment
  additional finance/investment 46–8
  Latin America 221–9
  public *see* public investment
  water industry 93, 95–106
  see also conventional public investment
Italy, electricity industry 128, 129, 130–1, 134, 189–90

Latin America
  capital formation 229–37
  foreign capital 215–22
  mergers & acquisitions 202, 220, 223, 225
  private investment 221–9
  privatization highlights 219
  privatization 202–45
  privatization revenues 214
  public finance 229–37
  Public Private Partnerships (PPP) 16
  public services 221–9
  regional/sectoral concentration 210–15
  restructuring 203, 210, 220, 233
  state-owned enterprises 204, 209, 213, 232, 238
  structural adjustment 204
  transnational corporations (TNCs) 222–9
  water industry 85–6, 90, 91, 94, 96–9, 101, 112

Lobina, Emanuele 75–120
London Underground
  PPP projects 3, 9–10, 12, 16, 17, 32
  privatization 31
Luxembourg, Rosa 30

mergers & acquisitions 202, 220, 223, 225
Ministry of Defence (MoD) 56–7
multinationals *see* transnational corporations

National Air Traffic Services (NATS) 9–10, 15–16, 60
National Audit Office (NAO) 6, 8, 10–12, 14–15, 17–18, 53, 65–6, 70
nationalization, working conditions 28
neo-liberalism
  globalization 27
  ideology 3
  New Right 1, 27, 33
  private finance 26–33
net present cost (NPC) 7, 8
New Public Management (NPM) 41
New Right, neo-liberalism 1, 27, 33

operational performance, penalties 17–18

partnerships
  PPP *see* Public Private Partnerships
  risk factor 8, 11, 60
  value for money (VFM) 13
Partnerships UK (PUK) 9
political economy, Private Finance Initiative (PFI) 1–38
pricing
  risk 56–9
  water industry 107–110
private finance
  costs 3, 54–5
  financial advisers 1, 3, 6, 10
  neo-liberalism 26–33
  profit motive 4
  wealth creation 4–5

Private Finance Initiative (PFI)
  accounting issues 48–52
  additional finance/investment 46–8
  affordability 10–11, 20
  Andersen report (2000) 12
  building to time/budget 14–15
  competitive pressure 4, 6
  conflict of interest 9–10, 25, 26
  contracts *see* contracts
  conventional public investment distinguished 40–1
  costs *see* costs
  discount rate 7, 8, 45, 51
  ex ante evidence 11–13
  ex post evidence 13–14
  financial advisers 3, 6, 10
  government bail-out 10, 16, 32–3, 56, 60
  Institute of Public Policy Research report (2001) 13
  penalties/incentives 17–18
  political economy 1–38
  privatization 41
  profit 4, 11, 21, 22
  programme control 6–24
  public services 2
  PWC report (2001) 12–13
  robust specification 15–16
  scale 43–6
  tenders/bids 6, 15, 69
  Treasury 8, 9, 10, 15, 25, 26
  VFM *see* value for money
private sector, secondees 9, 24
privatization
  background 166–7
  debates 122–6
  electricity industry 122–6, 160–201
  Latin America 202–45
  London Underground 31
  policy promotion 24–6
  Private Finance Initiative (PFI) 41
  targets 30–1
  United Kingdom 204–5
  United States 205–6
  water industry 75–120
  World Bank 206–7
profit
  capital employed 28

Private Finance Initiative (PFI) 4, 11, 21, 22
profit motive 4
social relations 27
surplus value 28
public accounts
  assets/liabilities 48–9
  balance sheets 49–50
Public Accounts Committee (PAC) 6, 53–4, 56–7, 60, 69–71
public borrowing
  interest 52, 54, 55, 56
  risk 52, 55, 56
public debt 49
public expenditure
  budget deficits 46, 48, 50
  golden rule 50
  investment *see* public investment
  tax *see* taxation
public investment
  capital expenditure 46, 50
  conventional *see* conventional public investment
  finance/funding 48
  opportunity cost 47
Public Private Partnerships (PPP)
  capital values 5, 21, 44
  contracts 67, 68
  definition 40
  health/hospital schemes 10, 15–17, 18–20, 23–4, 62
  insolvencies 10, 16, 19
  IT projects 12, 18, 23
  joint ventures 2
  Latin America 16
  London Underground 3, 9–10, 12, 16, 17, 32
  National Air Traffic Services (NATS) 9–10, 15–16, 60
  prison schemes 17–18
  transport schemes 3, 5, 9–10, 12, 14–17, 20–1, 24
public sector
  risk 62
  water industry 76–80
public sector comparator (PSC)
  appraisal methodology 7, 8, 12, 42
  comparison skewed 10, 64, 66
  costs 7, 42, 56–7, 63, 64, 65, 72

Index

discount rate 65
front loading 65
risk 7, 56–7, 63, 65
public services
  Latin America 221–9
  Private Finance Initiative (PFI) 2
  transnational corporations (TNCs) 26, 27

refinancing 16, 19, 22, 53
risk
  basic argument 55–6
  control 55
  conventional public investment 54–5
  default risk 52, 55, 56
  effective transfer 59–61
  ex ante 58
  net change in amount 61–3
  partnerships 8, 11, 60
  premium 7–8, 11, 20, 21, 54, 60
  pricing 56–9
  probabilities 61
  public borrowing 52, 55, 56
  public sector 62
  sharing 40, 54–63
  transfer 21–3, 39, 43, 54, 56–61, 63, 72
  value for money (VFM) 7, 8, 13
  variability risk 55
Robinson, Geoffrey 10

Sawyer, Malcolm 39–74
Shaoul, Jean 1–38, 60–1, 62–3
shareholder returns, post tax return 20, 21
socialism, political consciousness 31
Spain, electricity industry 128, 129, 130, 134, 184–7
special purpose vehicles (SPV) 20, 22
Sweden, electricity industry 128, 129, 135, 136

taxation
  corporate taxation 28–9
  debt payment 55
  post tax return 20, 21, 22
Tequila Effect 207

transnational corporations (TNCs)
  globalization 29
  Latin America 222–9
  public services 26, 27
  water industry 83–4
Treasury 8, 9, 10, 15, 25, 26

United Kingdom
  electricity industry 128, 129, 132–3, 182–4
  PFI *see* Private Finance Initiative
  privatization 204–5
  water industry 81, 82–3, 86–7, 102–4, 108–10, 113–14
United Nations, Millennium Development Goals (MDGs) 95
United States, privatization 205–6

value for money (VFM)
  appraisal methodology 7–9
  competition 6–7
  costs *see* costs
  partnerships 13
  Private Finance Initiative (PFI) 3–4, 6–10, 12, 13, 41–2, 50
  risk factor 7, 8, 13
Vidal, Gregorio 202–45

wages, reductions 3, 18
water industry
  Africa 95, 112
  Asia 95, 112–13
  Buenos Aires 96–7
  Build-Operate-Transfer (BOT) 99–100
  Chile 97–8
  Colombia 98–9
  comparisons 101
  contracts 85
  corruption 84, 86
  developing countries 79–80, 84, 115
  efficiency 110–14
  European Union (EU) 77, 104–6, 109
  expansion 81–4
  failures 87–8
  France 77–9, 81, 82, 107–8

252 Index

water industry – *continued*
  gaming under regulation 108–10
  history 76–80, 101
  investment 93, 95–106
  Latin America 85–6, 90, 91, 94, 96–9, 101, 112
  multinational groups 83–4
  OFWAT 102–4, 109–10
  ownership 91–2
  political movements 85–7
  pricing 107–10
  privatization 75–120
  problem areas 84–8
  public sector 76–80
  public/private comparison 111–14
  renationalization 94
  research and development (R&D) 114
  retreat 88–9
  solidarity finance 104–6
  south 90, 93
  state capture 86
  United Kingdom 81, 82–3, 86–7, 102–4, 108–10, 113–14
  water companies 81–93
  World Bank 93, 98, 111, 115
World Bank
  electricity industry 123–4
  privatization 206–7
  water industry 93, 98, 111, 115